# TASTE BUDS AND MOLECULES

# TASTE BUDS
## AND MOLECULES

## THE ART AND SCIENCE OF FOOD WITH WINE

## FRANÇOIS CHARTIER

Translated by Levi Reiss

McCLELLAND & STEWART

**Library and Archives Canada Cataloguing in Publication**

Chartier, François
    Taste buds and molecules : the art and science of food with wine /
François Chartier ; translated by Levi Reiss.

ISBN 978-0-7710-2267-8

1. Food—Sensory evaluation.  2. Taste—Molecular aspects.
3. Food—Composition.  4. Wine tasting.  5. Wine service.
6. Gastronomy.  I. Reiss, Levi  II. Title.

TX546.C4213 2011          664'.072          C2011-902105-6

We acknowledge the financial support of the Government of Canada through the Book
Publishing Industry Development Program and that of the Government of Ontario through
the Ontario Media Development Corporation's Ontario Book Initiative. We further
acknowledge the support of the Canada Council for the Arts and the Ontario Arts Council
for our publishing program.

Graphic design
www.cyclonedesign.ca

Photography
Michel Bodson, Studio F2.8 (pages 46, 48, 53, 58, 64, 68, 109, 115, 127, 147, and 175)

Culinary stylist and chef
Véronique Gagnon-Lalanne (pages 46, 48, 53, 58, 64, 68, 109, 115, 127, 147, and 175)

Typeset in Trade Gothic by M&S, Toronto
Printed and bound in China

McClelland & Stewart Ltd.
75 Sherbourne Street
Toronto, Ontario
M5A 2P9
www.mcclelland.com

1   2   3   4   5         14   13   12   11   10

# TABLE OF CONTENTS

# FOREWORD

JULI SOLER AND FERRAN ADRIÀ

When we met François Chartier several years ago, we quickly realized that we were in the presence of an exceptional talent. He has an incredible vivacity and possesses such a dynamic and supple thinking process that if someone had told us we would soon collaborate in our kitchen, as is now the case, we would not have been very surprised. François also has a very vigorous work methodology, as well as a profound knowledge of the world of wine, his mastery of which allows us to qualify him as the number one expert on flavors.

Using his insightful research on wine and his unflagging search for the secrets of successful wine and food pairings, François is now here to help us refine the information conveyed by our senses and understand why some wine and food pairings work better than others. The answer to this question is found in the molecules that certain of these pairings share. Chefs and sommeliers do not traditionally have this kind of scientific information at their disposal, even if they have been creating successful wine and food combinations on an intuitive level.

As a result of his innovative research, François has unearthed new, revolutionary food and wine pairings that are just as felicitous as those suggested by tradition. These surprising discoveries will satisfy chefs, sommeliers, and diners alike. To cite one example, sommeliers usually suggest Mediterranean wines to accompany dishes prepared with rosemary. François Chartier has identified various aromatic molecules common to rosemary and some Alsatian wines. He has thus created an innovative approach, an original pairing that makes sense from a gastronomic point of view for one very simple reason: it works! In other words, gustatory intuition, a tool of chefs and sommeliers from time immemorial, confirms and proves the success of this new wine and food pairing.

There will always be those who say that we cannot reduce the magic of wine, of cuisine, of flavors in general, to simple chemical formulas. And it's true that eating and drinking are much more than all that. But knowledge is one of our most important tools, and the more we master all aspects of the raw material with which we work on a daily basis, the better we can satisfy our attempts at creativity, and deliver the pleasure that our clients expect. François likes to make the following comparison: a musician can be truly happy while playing his instrument, and do so extremely well, even if he has no theoretical knowledge. If subsequently he acquires such understanding, his music will only get better and better. This new knowledge will by no means diminish the pleasure, the spontaneity, the mastery, and the creativity of his art. The art of cooking works the same way. Greater wisdom and increased knowledge generate additional possibilities to create, to evolve, and to satisfy.

It's for all these reasons that we enthusiastically suggest that you let yourself be charmed by this magnificent book, and step into the magic of new flavors. Allow yourself to be seduced by your taste buds and by the molecules that François Chartier explains, ever so clearly, in these pages. We are convinced that this is a groundbreaking work, a first step into a world which from this moment on will splendidly unfold for all those who love gastronomy.

**Juli Soler and Ferran Adrià**
Co-owner and chef/co-owner of the famous elBulli restaurant in Roses, Spain, five times named "Best Restaurant in the World"

# DR. RICHARD BÉLIVEAU
# FOREWORD

This book is an essential work for all wine lovers and for all those who love the pleasures of the table. It presents the results of a truly scientific approach applied to understanding the molecular mechanisms that govern gastronomic pleasures. The work of François Chartier transports us into a fascinating world, one inhabited by complex aromatic molecules with enchanting names. These molecules not only act upon multiple biological processes associated with maintaining physiological equilibrium (homeostasis), but also, by a strange quirk of fate, cooperate among themselves to stimulate those cells of ours that specialize in detecting flavors and providing pleasure. Whether it be eugenol, cinnamic aldehyde, coumarin, capsaicin, or any other molecule of vegetable origin, all these compounds have the important characteristic of simultaneously activating our flavor detection systems while positively influencing the mechanisms that contribute to the homeostasis responsible for maintaining our health.

Understanding the molecular mechanisms underlying gastronomic pleasure is therefore not only essential for more harmoniously pairing the flavors in our daily meals; this increased understanding also provides an important step in improving our diet's impact on our health.

Though it may be heir to thousands of years of empiricism in the service of identifying the best food combinations, current gastronomic knowledge is not a static, immobile science whose limits are etched in stone. The courageous, pioneering work of François Chartier illustrates to which point a modern, dynamic approach applying current scientific knowledge can innovate and identify new wine and food pairings and constantly redefine the limits of pleasure. About 200 years ago, the French gourmet Brillat-Savarin said, "Taste, as nature bequeathed us, is still the sense that, when you consider the whole picture, provides us with the greatest pleasure."

Thanks to the work of François Chartier, we finally can start to understand why.

**Richard Béliveau, Ph.D.**
Cancer researcher; Claude-Bertrand Chair in Neurosurgery (Centre hospitalier de l'Université de Montréal); Chair of Cancer Prevention and Treatment/biochemistry professor (Université du Québec à Montréal); researcher, Oncology Services (Montreal Jewish General Hospital); and co-author of three books, including *Cooking with Foods That Fight Cancer*.

# ACKNOWLEDGMENTS

I want to emphasize the precious collaboration of my gastronomic and scientific friend Martin Loignon, who helped with the scientific editing of several chapters of this book. Martin, who holds a doctorate from the University of Montreal in molecular biology and has a great deal of experience in the field, is presently a senior scientist for a large Montreal research firm. Having been inspired for a long time by cuisine and by wine and food harmony, his twofold knowledge of gastronomy and science made him an ideal resource, able to peer knowledgeably over my shoulder. In addition to his work on several other chapters, he is the author of the chapter "The Culinary Revolution."

A heartfelt thank you also goes to Richard Béliveau, Ph.D., distinguished professor, cancer researcher, and author. He proved to be the first scientist to support me when I had the idea to research wine and food pairings as related to molecular sommellerie. Meeting with him would prove to be the deciding factor in my decision to pursue this path. That he wrote one of the two forewords to this book turned out to be "the icing molecule on the cake"!

Un fuerte abrazo (a big hug) to Ferran Adrià and Juli Soler, from the Catalan restaurant elBulli, whose innovative work has been transforming my way of thinking since 1994, the year that I discovered the existence of these enlightened creators with whom I am now working (for more details, see the chapter on elBulli). The fact that they affixed their signature on one of the two forewords of Taste Buds and Molecules deeply moved me.

I also want to point out the inspiring efforts of Pascal Chatonnet, a reputed Bordeaux oenologist and co-owner of the Excell professional wine laboratory and of several vineyard chateaus in Lalande de Pomeral, France—including Château Haut-Chaigneau and Château La Sergue in Libourne, as well as Château L'Archange in Saint-Émilion—whose two doctoral theses on the aromatic impact of oak wine barrels during wine maturation profoundly influenced me. I greatly benefited from his collaboration and support for my project on wine and food pairing research (see the chapter "Experiments in Food Harmony and Molecular Sommellerie"), as well as his careful reading of certain parts of this book. Finally, he was one of the first two recognized specialists to confirm to me that I absolutely had to explore in depth this molecular path.

More than friendly thanks go to Chef Stéphane Modat, former master chef of the restaurant L'Utopie in Quebec City, with whom I was able to present my first practical examples of food harmony and molecular sommellerie. This was in October 2008, during the launches of my book La Sélection Chartier 2009, and in March 2009, during our "Molecular Tasting Meal with Two Master Chefs" (see the chapter "Experiments in Food Harmony and Molecular Sommellerie"). Each of our meetings was like a seismic explosion in which innumerable creations

013

gushed forth. I hope there will be many other "creative earth-quakes" to come!

I raise my hat to Hervé This for his creations, which have been inspiring me since the mid-1990s, and also for his communicative energy and the many ideas that we shared during our 2007 meeting.

A tip of the hat also to Josep Roca, renowned sommelier and co-owner, with his two brothers, chefs Joan and Jordi, of the great restaurant El Celler de Can Roca, in Girona, Catalonia, Spain. He has created a unique wine cellar, the work of a lifetime, that all self respecting sommeliers should visit at least once during their careers. Thank you, Josep, for your openness, for the inspiration you gave me, and for the confidence you had in my work.

Major thanks to Nicole Henri for having agreed to plunge into the domain of molecular chemistry, even though it was not her world, in order to edit my texts, as she has done so well for the past several years. Equal thanks to Martine Pelletier and Martin Balthazar from Éditions La Presse, who, besides carefully managing this project, respectfully enlightened me during the rereading of the final proofs of this work.

My thanks to the team of Vinum Design (www.vinumdesign.com), who furnished us with the glasses and carafes used in certain photos in this book.

And finally, a thank you from the bottom of my heart goes to Carole Salicco, my wife and constant collaborator who, once again, supported me during this crazy wine and food pairing adventure, according me complete freedom to consecrate myself entirely and then some (!) to the project. Without you, Carole, the volatile compounds of daily life would not have so much flavor…

# AUTHOR INTRODUCTION

In this work, I am introducing the first results of my scientific and gourmet research on wine and food pairing and molecular sommellerie. My goal is to try humbly to bring new light to wine and food pairing by exploring the aromatic path of foods, wines, and other beverages.

Since 2006, after twenty years of experiments, I have been working to "map out" the aromatic molecules that give foods and wines their taste. This finely detailed work at the heart of a new scientific domain has enabled me to better identify the bridges that exist between the foods you eat and your favorite wines.

To my great surprise, the first results of these wine and food pairing adventures also enabled me to acquire a richer and more precise knowledge of foods' molecular identities. This led me to make connections, sometimes fairly surprising ones, among certain complementary ingredients. The result was an array of new possibilities for harmonious food combinations and new paths to culinary creativity, both for novice cooks and professional chefs.

Supported by an imposing body of scientific literature and inspired by my priceless collaborations with chefs, oenologists, and scientists from the world of food and wine, both in Quebec and elsewhere, I was able to identify the principle volatile compounds that impart their aromatic and gastronomic identities to several foods and wines. This work enabled me to explain how a molecular relationship between two foods, or

between a given wine and a food, guarantees a harmonious pairing.

This book is a bit like a Polaroid in that it reveals my present progress in the enterprise of tracking down aromatic secrets, but it is very much an ongoing process.

It is also my aim in this book to suggest simple and easy-to-use hints and ideas, both for day-to-day cuisine and for festive meals. You have in your hands a book that is practical, abundantly illustrated, adorned with photos, sketches, and drawings, and filled with recipes for daily use, tricks for sommelier-cooks, and even cocktail recipes—not to mention a number of recipes from great chefs.

As you may well imagine, the food and wine "trees" that I have been sketching for the past twenty years while on the trail of harmonious wine and food pairings are a work in progress. They have become increasingly complex, transforming according to the growing knowledge of molecular principles that I have been intensively building since 2006.

In this book, I wish to, in a sense, "set the table" for you: to let you practice your aromatic scales in this new field and refine your gourmet tastes. My hope is that future volumes, in which other foods and wines will be added to the menu, will allow you to take this knowledge yet further.

I hope that once you have finished reading each chapter, you'll be inspired by the drawings of foods and complementary wines to unearth your favorite recipes based on these ingredients, or to create others yourself—and to pair them harmoniously with the requisite wine!

Once again, thank you for joining me for this rendezvous, and I'll see you very soon for new adventures in wine and food pairing.

**François Chartier**

P.S. Does reading this book raise questions about food, wine, or wine and food pairing? Ask questions at my website, **www.tastebudsandmolecules.com.** I will try to answer them to the best of my ability.

MARVIN

CHARTIER LOIGA

FERRAN ADRI

ULLI SOMMELER

JULI SOL

CINNAMON
SCHNAPPS WITH
GOLD FLAKES

AN EL BULLI ALVARO

GOLDSCHLÄGER

PALACI

JOSEP ROC

CHARTREUSE

PEIRCE MOLECULAR

ENTELLES SANTA

ADRIA

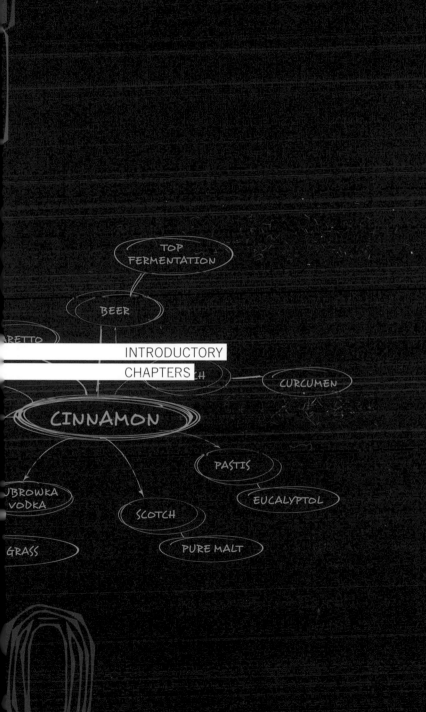

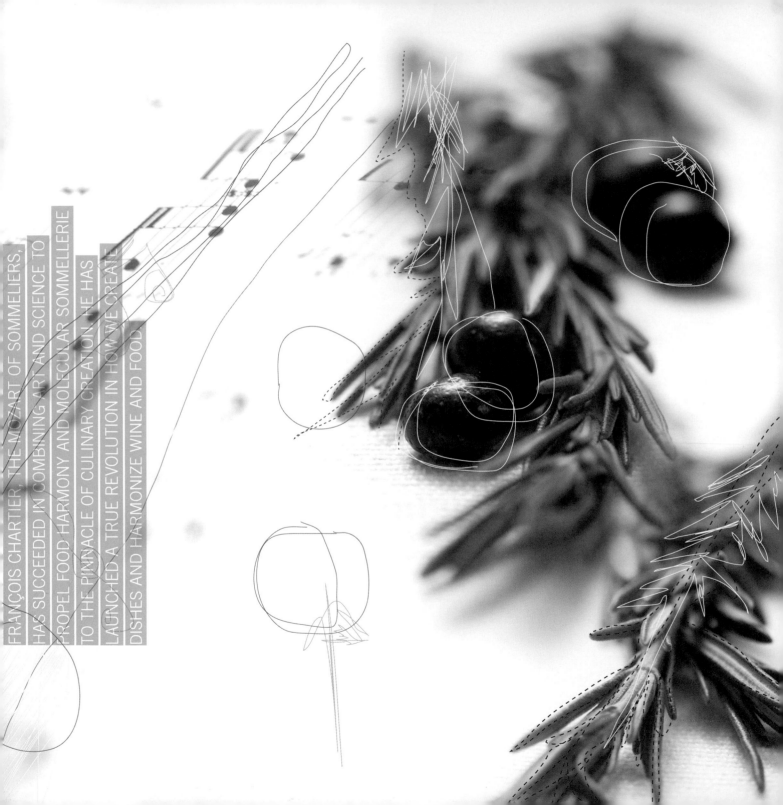

FRANÇOIS CHARTIER, THE MOZART OF SOMMELIERS, HAS SUCCEEDED IN COMBINING ART AND SCIENCE TO PROPEL FOOD HARMONY AND MOLECULAR SOMMELLERIE TO THE PINNACLE OF CULINARY CREATION. HE HAS LAUNCHED A TRUE REVOLUTION IN HOW WE CREATE DISHES AND HARMONIZE WINE AND FOOD.

# THE CULINARY REVOLUTION

## UNLEASHED BY THE PRINCIPLES OF FOOD HARMONY AND MOLECULAR SOMMELLERIE

By Martin Loignon, Ph.D.,
molecular biologist, Montreal.

Throughout our existence, humankind has created and modernized musical instruments and mastered the tones and sounds that they produce. In their respective heydays, Vivaldi, Mozart, Tchaikovsky, and Gershwin knew how to take advantage of these innovations and succeeded in integrating new sounds into their oeuvres. Swept up by creativity, enlightened by profound knowledge of composition, in collaboration with violin makers, artisans, and musicians, they constructed scores that guaranteed the timelessness of their works. The greatest conductors are those who have mastered every note of their scores, and who direct their musicians to vary the music's intensity at specified moments to create a memorable musical experience for their audience.

Scores and notes are to music as foods, wines, and aromatic molecules are to gastronomy. Their power to unleash sensory pleasures essentially depends on the chef's orchestration. In much the same way, chefs and composers harmonize scores that effectively consist of ingredients. When painstakingly assembled, properly heated, and expertly enhanced, these ingredients will arouse pleasures stemming from their aromas and textures, all the while awakening sensory memories. A knowledgeable sommelier, master of his art, will, by precisely pairing wines and foods, enhance the olfactory pleasures of the dishes and heighten the culinary memories of the diners at his table.

Unlike with music, humankind has not succeeded in creating foods from which one can identify individual aromatic notes. At best, we can transform them, discovering and recognizing both naturally occurring and transformed aromas, all the better to harmonize them. In contrast to symphonies, food pairing has always depended more on the chef's instincts than on her knowledge of foods' aromatic molecules. Like a musician who ignores a score's notes and plays by ear, the chef harmonizes by "nose." This is because even though we have a relatively wide understanding of the aromatic molecules (the notes) that make up foods and beverages, this knowledge is in fact virtually unknown, unexploited, and even inaccessible to all but a few. This leads to uncertainty about which pairings are possible and fosters the fear of innovation. It's not for lack of creativity but rather for lack of knowledge that today's chefs, even the greatest, succumb to the temptation of repeatedly pairing the same ingredients.

The most complicated feat in cuisine may be to do it simply. But how can one do things simply in a field that still hides so many secrets? The cuisine of the future, like that of the past, will please and seduce us by its innovations. New instruments and new methods of using the familiar ones, but above all a heightened understanding of foods, ingredients, beverages, and wines, will open the doors to new food and wine harmonies and pairings. In this respect, the new field known as food harmony and molecular sommellerie has, in the last few years, increased our knowledge of these harmonies more than any other science.

Revolutionary ideas and innovations contribute to changing viewpoints, mentalities, and ways of doing things. Witnessing a revolution, even if one doesn't fully understand its impact, creates a feeling of experiencing a unique, historical moment. One becomes quite privileged as soon as one understands the fundamentals underlying innovation because it then becomes possible to evolve from spectator to participant. The arts and sciences have always been fertile grounds for innovation. Several major innovations can be found at the crossroads of art and science; architecture, cinema, music, and, more recently, research on molecular gastronomy were all born from the fusion of art and science. The meeting of these two seemingly separate universes has become a major fuel for contemporary creative minds.

In this monumental work, which will profoundly change the culinary arts, Chartier unveils secrets of foods, ingredients, beverages, and wines that have remained unknown for far too long. He does so passionately and convincingly, backed up by several years of research and exemplary scientific rigor. Not only does he allow us to discover the molecules responsible for

FRANÇOIS CHARTIER, THE MOZART OF SOMMELIERS, HAS SUCCEEDED IN COMBINING ART AND SCIENCE TO PROPEL FOOD HARMONY AND MOLECULAR SOMMELLERIE TO THE PINNACLE OF CULINARY CREATION. HE HAS LAUNCHED A TRUE REVOLUTION IN HOW WE CREATE DISHES AND HARMONIZE WINE AND FOOD.

aromas; he offers, as if by magic, an abundance of innovative wine and food pairings that truly transform theory into practice.

This concept is perfectly illustrated in the chapter "Fino and Oloroso," which discusses sherry, that misunderstood libation that is capable of holding its own with the greatest wines and sports a conviviality that would make many a Chablis blush and turn the best Côte-Rôties green with envy. Sherry, as Chartier explains, stands up readily at the table with a multitude of exotic dishes and harmonizes with each of them.

Sherry, the chameleon of wines, has molecular attributes that subtly mesh with the aromas of the most delicate dishes, while still being able to enhance neutral dishes and not fade away when paired with spicy ones. The secret of sherry's versatility, which instinctively one might explain by its flexibility, its simplicity, or in a negative sense by its lack of character, is actually something else entirely. In fact, the versatility of sherry is due to its aromatic complexity, the result of hundreds of molecules (more than 300) associated with walnuts, caramel, butter, apple, apricot, and many others. Given this great number of aromas, there is a good chance that sherry will find its alter ego in a wide variety of ingredients and foods. In addition to this aromatic multiplicity, a judicious dose of strength and sweetness favors the consonance between molecules with aromatic affinities, and discourages discordance and clashing due to molecular incompatibilities.

Does molecular wine and food pairing frighten you? Take up the rhythm by making sherry the metronome of your experiments. Without it, you may easily come up with the couplets of a meal, but you'll always miss the refrain.

If there were a Nobel Prize for gastronomy, François Chartier would be a deserving recipient. It took a large measure of genius, creativity, and audacity to construct the foundations and rules establishing the cause and effect relationship between aromatic molecules in foods and wines, and linking them together to create successful pairings. One could very roughly summarize this work as the combining of foods, ingredients, wines, and beverages that contain the same aromatic molecules so that harmonious pairings may occur spontaneously, with aromatic compounds seeking similar ones out. But this is an extreme oversimplification of this tremendous undertaking. To truly perfect this art and science requires a knowledge of the aromatic signature of each and every ingredient in a dish and the accompanying wine, in addition to an understanding of all the subtleties that occur when elements interact according to the author's principles of harmony. Molecular gastronomy will, henceforth, take its cues from food harmony and molecular sommellerie.

Please note that this book is not addressed only to culinary and wine professionals, or to experts in chemistry. It's also for the curious, and in particular for anyone who wants to open a window into a world of infinite possibilities, to achieve culinary success in creating food pairings that are innovative and unique, yet still harmonious. It is also addressed to all those who are ready to participate in this culinary revolution, to those for whom the pleasure of eating well is essential. This book will give both the neophyte cook and the most experienced chef the confidence necessary to transgress culinary traditions and forge new paths by trying out wine and food pairings that seem unlikely at first glance.

WE MUST UNDERSTAND THE
CHEMICAL REACTIONS THAT GOVERN
WINE AND FOOD PAIRINGS

I AM RESEARCHING "VOLATILE
MOLECULES" AND CHARTING IN
DETAIL THE AROMATIC COMPOUNDS
OF FOOD AND WINE

SCIENCE AND WINE. SCIENCE AND CUISINE
SCIENCE AND WINE AND FOOD PAIRINGS!
SCIENCE AND WINE AND FOOD PAIRINGS!

# FOOD HARMONY AND MOLECULAR SOMMELLERIE

## THE GENESIS OF A NEW HARMONIOUS SCIENCE BALANCING CUISINE, WINE, AND WINE AND FOOD PAIRINGS

"The process determines the product. If you don't change the process, you will always end up with the same product."

FRANCO DRAGONE, DIRECTOR, CIRQUE DU SOLEIL

### SCIENCE AND WINE, SCIENCE AND CUISINE... SCIENCE AND WINE AND FOOD PAIRINGS!

Since the end of the 1980s, wine growers have benefited from scientific advances in winemaking in order to better understand their profession, which had often been learned empirically. At the four corners of the earth, even in those regions where scarcely twenty years ago it would have been unthinkable to cultivate wine grapes profitably, winemakers are now producing the best wines in history.

Since the mid-1990s, the results of scientific research on molecular gastronomy (a discipline dating back to the early 1980s), first conducted in laboratories and then adapted to the kitchen, have provided cooks and chefs with a more precise, more scientific understanding of the empirical procedures they had been using, which had originated in century-old cookbooks.

All that was missing was a single link in the molecular chain: understanding the chemical reactions that govern wine and food pairing. This required a scientific understanding of the molecular structures of different foods.

After three years of somewhat hazy reflections, from 2003 to 2006, a new research path presented itself to me. I called this path "food harmony and molecular sommellerie."

"The upheavals of modern science have transformed our understanding of what knowledge is. Scientific knowledge is, from a practical point of view, the most exact and the most useful knowledge that we humans possess. Therefore, to be credible, every notion describing knowledge must also apply to scientific knowledge."

ALBERT EINSTEIN

FRANÇOIS CHARTIER SHOWING SOME OF HIS SKETCHES

## LOOKING FOR FOOD PARTICLES

Since then, I have been researching "volatile molecules" and charting in detail foods' aromatic compounds—starting with my "bridge ingredients," as described in my book *À table avec François Chartier*. I established the potential correlations among wine and food components with the goal of creating more precise wine and food pairings and opening new pairing vistas.

To reach my goal, I held discussions, exploratory meetings, and collaborations both in Quebec—with, among others, Dr. Richard Béliveau and the Food Research and Development Centre of Agriculture and Agri-Food Canada—and in Europe, with the Bordeaux wine scientist Pascal Chatonnet, chef Ferran Adrià of the Catalan restaurant elBulli, and members of the food research center Alícia, located near Barcelona.

In addition, I had several exchanges with food science and molecular biology researchers, including Martin Loignon, Ph.D., as well as with wine scientists and leading innovative chefs. Every single day, I put my nose to the grindstone, working on this project with the help of the vast body of scientific literature on molecular biology.

I have, to date, charted the molecular structure of a number of foods and wines. This has allowed me to propose, in this volume, new ways to create recipes for both cooks and chefs, and a new understanding of wine and food pairings for both wine amateurs and wine professionals. This work is meant to open to everyone the possibility of creating recipes based on harmony among components, as well as recipes specially conceived "by and for" the accompanying wine, in which all ingredients comprising a given recipe are in harmony with that wine.

Figs, vanilla, maple syrup, rosemary, saffron, mint, basil, root vegetables and all foods that taste of anise, cloves, cinnamon, strawberries, pineapple, tomatoes, seaweed, sesame seeds, ginger, licorice, different kinds of rice, coconut, mushrooms, lamb, beef, pork, even green teas and smoky black teas, to name only a few ingredients, unfolded their secrets to me and presented new paths to wine and food pairings.

I can say the same thing for red wines based on Cabernet Sauvignon, Merlot, Cabernet Franc, Syrah, Mencia, Tempranillo, and Grenache, as well as for white wines, those based on Muscat, Gewürztraminer, Scheurebe, Pinot Gris, Riesling, Sauvignon Blanc—and let's not forget sherries, whether fino, amontillado, or oloroso, and certain naturally sweet wines.

## SCIENCE: AN END IN ITSELF?

**"Over the course of the twentieth century, science has undergone profound changes. This has led, among other consequences, to the idea that our most advanced knowledge is merely made up of fallible theories subject to modification. Sooner or later, today's theories will be replaced by better ones. Human knowledge is fallible precisely because it comes from human beings; and now we are confronted with this discovery: science does not consist of unchanging certainties."**
ALBERT EINSTEIN

## WHAT PATH LED ME HERE?

+ I spent twenty years developing wine and food pairings by redefining from the very beginning what had been considered "absolute references" (1989-1992).

+ I started by selecting the wine, and then chose the food— we can't "change" the wine, but it is easy to adapt the cuisine to the wine (1992-1998).

+ After several years of selecting the wine before determining the menu and adapting the menu to the wine, and after having traveled and studied cuisine around the world, I was inspired by the aromas and structures of wine to create "made-to-measure" meals. This meant developing new recipes in order to achieve perfect pairings (1998-2002).

My "laboratory" during this fruitful period, which spanned from the late twentieth century to the onset of the twenty-first century, was the series of wine-tasting dinners that were part of the *Club de Vins François Chartier* (whose menus appear in the PDF document found in the *Club de vins* section of the website www.francoischartier.ca). This series was developed in collaboration with many talented chefs who were open to the spirit of Quebec. I want to take this opportunity to thank the chefs Bassoul from the restaurant Anise, Bastien from Leméac (especially for his

CHARTIER IN THE KITCHEN OF ELBULLI, ACCOMPANIED BY JULI SOLER AND FERRAN ADRIÀ, AS WELL AS JOSEP ROCA, SOMMELIER OF THE RESTAURANT EL CELLER DE CAN ROCA

collaboration during the first two *Montréal Passion Vin* galas), Besson from Laloux, de Canck from La Chronique, Desjardins from l'Eau à la Bouche, Fradeau from the Hotel Vogue, Gaildraud and Halbig from the Bistro à Champlain, Godbout from Chez l'Épicier, Laloux from the caterers Laloux, Laprise from Toqué!, Lemieux from Bouchon de Liège, Picard from Le Club des Pins (now with Au Pied de Cochon), Massenavette from La Clef des Champs, Claude Pelletier from Mediterraneo (now with Le Club Chasse et Pêche), and Tavares from the Ferreira Café. I also want to note my collaboration with two great French chefs on their visits to Quebec: Paul Bocuse from Bocuse and Philippe Legendre from Taillevent.

+ I rapidly discovered that certain ingredients, which I named "bridge ingredients," were the most important catalysts in successful wine and food pairings. So I redirected my research towards these bridge ingredients (2002-2006).

Ever since 2006, one thing has been certain: given the omnipresence of creative, avant-garde cuisine influenced by molecular gastronomy research—thanks to the joint efforts of scientists and chefs, including Ferran Adrià of elBulli—we absolutely must redefine wine's place at the table. In fact, this new cuisine will considerably influence our food preparation methods and how we eat over the course of the next fifteen years.

+ Since the spring of 2006, I have entirely devoted myself to this redefinition of wine's relationship to twenty-first century avant-garde cuisine. For example, I spent several weeks traveling during the winter of 2007, largely in Catalonia.

+ At the same time (2006), after several months of reflection, I integrated scientific findings into my work in order to better understand the wine and food pairings that I had been applying for the last twenty years. I did so especially to identify why certain ingredients become the harmonious bridges to some kinds of wines. This gave birth to the field that I call food harmony and molecular sommellerie.

+ Since then, thanks to the scientific literature and my collaborations with scientists, I have been able to discover the principal aromatic molecules that define the structural identity of foods and wines. This has allowed me to determine why some pairings work so well and to create new harmonious bridges, both on the plate and between plate and glass.

## THE PRAGMATISM OF C. S. PEIRCE

Another theory, expressed by the American mathematician and physicist Charles Sanders Peirce—who set out the basics of "pragmatism" with his ideas on knowledge as a form of practical intervention—was also part of my decision to apply science in the understanding of harmonious wine and food pairing.

In his first important work, *How to Make Our Ideas Clear* (1878), Peirce proposed that to understand a term clearly, we must ask ourselves what change its application would bring to our evaluation of a problematic situation, or to a proposed solution. This change constitutes the term's meaning. A term whose use changes nothing, he argued, has no verifiable meaning. Thus, Peirce presents pragmatism as a way to verify terms' meanings and consequently as a theory of meaning.

Peirce went on to say that knowledge is an active entity, and that we learn better when we treat knowledge as a practical activity. Furthermore, the questions of meaning and truth are best understood in this context.

According to Peirce, we acquire our knowledge as participants, and not as mere spectators.

In conclusion, he wrote, knowledge is a tool, perhaps the most important tool for survival that we possess. Its most useful aspect lies in its power to explain. We only rely on this power when, as for any explanation, it leads to accurate results. If we run into difficulties, we try to improve the theory, and perhaps even replace it. This means that scientific knowledge is not a collection of certainties, but explanations. Enhancing scientific knowledge does not mean adding new certainties to previously existing ones, but replacing present explanations with better ones.

## CONTINUING RESEARCH IN FOOD HARMONY AND MOLECULAR SOMMELLERIE?

I would certainly need a good twenty years to survey all the foods and wines that are found on our table.

I will continue my discussions, my collaborations, and my exploratory meetings, in Quebec and abroad, with researchers, wine scientists, chemists, and chefs—in particular with Juli Soler and Ferran Adrià of elBulli, where I worked as a pairing consultant, as well as with the Fundació Alícia center for scientific studies in gastronomy.

And, certainly, I will work to prepare a second volume, extending this initial volume to share my further research findings on food harmony and molecular sommellerie—all for the enhanced pleasure of your taste buds!

**François Chartier**

"There's only one country in Europe that could produce Ferran Adrià and Álvaro Palacios."

DECANTER MAGAZINE

IT HAS BEEN NAMED BEST RESTAURANT IN THE WORLD ON FIVE OCCASIONS.

El Bulli

# elBULLI

## A TRIP INTO THE UNIVERSE OF THE "BEST RESTAURANT IN THE WORLD"

> "An idea is a point of departure and no more.
> As soon as you elaborate it, it becomes transformed by thought."
>
> PICASSO

An homage to Ferran Adrià and Juli Soler, founders and co-owners of the Catalan restaurant elBulli. Along with their team of inspired creators, they have rewritten the modern history of cuisine and have transformed my approach to food, wine, and wine and food pairing.

It would be unthinkable for me to publish this book without thanking the elBulli team, who opened the doors to their workshop elBullitaller in Barcelona (what an inspiration!) and to their mythical restaurant in Roses.

Since 1994, I have become interested in their research through the groundbreaking findings they have published over the years on their website, www.elbulli.com. Their working methods, both in the kitchen and in the dining room, have transformed my way of working and my knowledge of food and cuisine, allowing me to launch the principles of food harmony and molecular sommellerie.

I will never be able to thank them enough for their generosity and open-mindedness, and for sharing their enormous knowledge, both in their books and in our meetings. They also, on a number of occasions, graciously demonstrated their confidence in my research on food harmony and molecular sommellerie. And on top of everything, Juli and Ferran honored me by writing a foreword to this book.

Since they have unveiled their secrets in several books, which is quite unusual for master chefs, I am hardly the only one to have profited from their research and work philosophy. Today, numerous chefs, cooks, and restaurateurs of all ages have seen their success magnified thanks to the elBulli approach.

### PRESENTING ELBULLI AND ITS PLAYERS

elBulli is included in the very select circle of three-star establishments listed in the famous *Michelin Guide*. It has been named Best Restaurant in the World five times (2002, 2006, 2007, 2008, and 2009) by *Restaurant Magazine*, the new British bible of international gastronomy, which each year designates the world's fifty best restaurants as chosen by more than 800 experts in the field.

It is well known that several Catalan, Basque, and Spanish chefs influence and in fact dominate international gastronomy, especially when it comes to creativity and innovation. They have done so since the end of the 1990s. The central figure in this renaissance is undeniably the world famous chef of elBulli, Ferran Adrià.

Chef Adrià is to cuisine what the Beatles were to popular music, what Bach was to classical music, and what Picasso was to painting and sculpture: simply a revolutionary. He has become a point of reference and will influence several generations of chefs and restaurateurs to come.

Juli Soler, who possesses a legendary charisma, is the restaurant's masterful chief executive and represents the name

FERRAN ADRIÀ BEING INTERVIEWED IN HIS BARCELONA WORKSHOP
BY CHRISTIAN RIOUX, JOURNALIST FOR *L'ACTUALITÉ* MAGAZINE.

INTERVIEW WITH JULI SOLER IN THE ELBULLITALLER

elBulli. He is the one-man band of the restaurant: entrepreneur, executive, director, maître d'hôtel, master of public relations, and sommelier. He's the one who built the remarkable wine list at elBulli. He propelled this establishment to the top of avant-garde gastronomy. When Ferran left the restaurant in 1982, Juli recognized his creative potential and brought him back the same year. What happened next is already history, but this history continues to unfold.

elBulli's 2009 menu set off a renewal in its very innovative history. Ferran Adrià is presently enjoying a new, fruitful period of intense creativity.

Having rubbed shoulders with him in his restaurant in September 2008 and in his creative workshop in December of the same year, I've noticed that there are always many new ideas jostling around in the head of this master of the taste buds. He is supported by a solid team of "chef-researchers," including his right hand, the great chef Oriol Castro, and the chef of new products, Eduard Xatruch.

Similar to Picasso who, after a series of creative periods (blue, pink, African…), launched a veritable revolution with the advent of cubism, then shook up his contemporaries with surrealism, the next creations signed "Adrià" will mark a new stage in international gastronomy!

## THE BIRTH OF A HARMONIOUS COLLABORATION

I had the pleasure of going to this famous establishment in 2006 to experience my first "elBulli festival" accompanied by Cristina and Álvaro Palacios. Álvaro is a great winemaker of the Priorat, Rioja, and Bierzo regions of Spain. He is a pioneer in the revival of the fine wines of Spain, both in that country and in the rest of the world.

> "There's only one country in Europe that could produce Ferran Adrià and Álvaro Palacios."
>
> DECANTER MAGAZINE

ÁLVARO PALACIOS AND FRANÇOIS CHARTIER IN THE ELBULLI KITCHEN, 2006

Since then, I have developed a relationship based on respect and friendship with Juli Soler, and later with Ferran Adrià, as well as with the elBulli kitchen and dining room teams, which include the sommeliers Ferran "Fredy" Centelles Santana and David Seijas. Our first exploratory meetings led me to present my findings on harmonious wine and food pairings in a workshop conference held in the giant food salon Alimentaria in Barcelona, in March 2008.

Juli Soler, who attended my presentation, invited me to share my findings with Ferran Adrià and their team of chefs and sommeliers right in the restaurant in September 2008. After three hours of presenting my wine and food pairing graphics (as provided in this book) and my new paths toward creative cuisine based on the principles of wine and food harmonies, I had constructive discussions with some fifteen master researchers of Catalan gastronomy. The great Catalan sommelier Josep Roca, from the exceptional restaurant El Celler de Can Roca (www.cellercanroca.com), was also invited for this occasion (for a commentary on this restaurant, see the "Fino and Oloroso" chapter).

Ferran Adrià spontaneously invited me back in December of the same year to spend a week with him and his team in their Barcelona kitchen workshop. This enabled me to move forward and to inspire some new concepts for elBulli's 2009 menu.

Once I started to consult the list of ingredients with which the elBulli team had been working for several weeks, wine and food pairing ideas started to jostle in my head. The ensuing

exchanges were fruitful, intense, and often immediate as we experimented with new ideas, products, recipes, and pairings.

The kitchen team asked me to share my findings by adding other ingredients that would pair harmoniously with those they had listed; the goal was to create a symphony of flavors in the dishes themselves, and then between the food and the wine.

In response to a question posed to him in March 2009 by a gastronomy journalist from Quebec—"How will François Chartier's findings be reflected in your restaurant's coming menu?"—Juli Soler answered: "First, because François works

with our chefs Oriol Castro, Eduard Xatruch, and Ferran Adrià to research foods and how to pair them with other foods and with wine. Of course, some recipes on our new menu will carry the mark of his labors."

Our collaboration has continued. We email each other about the concepts we have shared and perfected and also about ongoing research projects. My return to elBulli in July 2009 enabled us to push the envelope further together. All of this culminated in a myriad of new creations on the 2009 menu.

WORKING ON HARMONIOUS PAIRINGS IN THE ELBULLITALLER
WORKSHOP WITH FERRAN ADRIÀ AND ORIOL CASTRO

CHARTIER EXPLAINS HIS RECENT RESEARCH TO JULI SOLER
AND FERRAN ADRIÀ IN DECEMBER 2008 UPON HIS ARRIVAL IN THE
ELBULLITALLER KITCHEN-WORKSHOP

Photos : Enrique Marcos, Barcelona

## THE SPIRIT AND SOUL OF ELBULLI'S 2006, 2007, AND 2008 MENUS

My wife and constant collaborator, Carole Salicco, and I had the privilege of experiencing what I call the "elBulli festival" in 2006, 2007, and 2008. We savored menus consisting of some 35 dishes each, showcasing a great avant-garde cuisine.

To sum up these menus that we were lucky enough to experience, here are the three words that my wife addressed to Juli Soler during the final meal of 2008: "air," "earth," and "water." These words described, respectively, the impressions left with us by the 2006, 2007, and 2008 menus. Juli Soler responded: "We are going to have to do 'fire' in 2009!" This isn't far from the truth, because of the immense creative transformation that animated Ferran Adrià in preparing the new 2009 season.

In 2006, the tasting menu at elBulli was studded with airy creations, of which several tapas/dishes were conceived by the process of spherification, with precise flavors that filled the mouth and persisted for an incredibly long time, although they were almost ethereal. In 2007, the approximately thirty-five tapas/dishes that marched across our palates were more earthy, more untamed, more "material," more strongly savory. The Catalan concept "earth and sea" was omnipresent, with a sharp earth accent.

Finally, in 2008, the spirit and soul of Japan, which had started to emerge in some of the 2006 and 2007 creations, became a stronger presence on the menu. Thus was presented a more Asiatic approach to cuisine, although without distorting the Catalan taste origins of Ferran Adrià. This time, the Catalan "earth and sea" concept was impregnated with more stress on the sea than on the earth, and thus more "water."

To get a more practical idea of elBulli's creations, I would encourage you to consult their various works, including the most recent, the very successful and accessible *A Day at elBulli* (Phaidon Press), as well as their remarkable website, www.elbulli.com.

# AROMAS AND FLAVORS

## THE FUNDAMENTAL IMPORTANCE OF AROMAS IN IDENTIFYING AND APPRECIATING THE TASTE OF FOODS, WINES, AND BEVERAGES

**"Aromas are one of the fundamental elements that contribute to a wine's sensory character."**
*REVUE DES OENOLOGUES*, APRIL 2006

Spices were at the heart of the great explorer Marco Polo's Asian expeditions in the twelfth century. The value granted these aromatic ingredients was already very high at that time. A pound of ginger cost a nanny goat. As for pepper, it was worth more than its weight in gold!

Even if a food's volatile molecules make up only 0.05% to 1% of its total molecular weight (the mass represented by its atoms), they are crucial in determining its taste. The molecular weight is the equivalent of a mole ($6.02 \times 10^{23}$) of molecules. For example, the molecular weight of water is 18 grams ($H_2O = 1 \times 2 + 16$).

According to researchers in nutritional science, between 80% and 90% of all sensations stimulating our appetite come from fragrances. Without these aromas, your morning bread and strawberry jam would be bland and tasteless!

### A PROOF: NEWTON'S APPLE

If you want to confirm the role that fragrances play in our diet, just bite into an apple while holding your nose. You will find that it is impossible to detect the apple's flavor and aroma. If you cover your eyes, you won't even know that it's an apple you're eating! You'll detect a tiny sensation of acidity and the fruit's texture, but you'll be missing the fundamental data identifying the apple's physical presence and aromatic molecules.

To continue this experiment, unplug your nose. The apple's taste will take over your olfactory bulb (a part of the brain involved in the perception of odors) and you'll say, "Eureka! It's an apple!" (or something to that effect). Without your nose, it is simply impossible to recognize what you're eating and drinking.

Newton would have really helped us understand the tasting process, in addition to setting out the fundamentals of the law of gravity, if, after the apple fell on his head, he had taken a bite while holding his nose!

So it's really the volatile molecules that, by their physical presence and aromas, make it possible for food and wine to have any taste.

According to the French magazine *Revue des oenologues,* "Aromas are one of the fundamental elements that contribute to a wine's sensory character. They occupy a major place in oenology [the science of wine and wine making], from the daily work in the vats to the wine's eventual consumption. Unfortunately, there is usually quite a lack of rigor when it comes to smelling wines. Major shortcomings range from the description of a wine's smells to professional training of wine smellers. This professional training is usually too cursory to be really complete and serious." (April 2006)

Aromas are based on compounds, including acids, alcohols, phenols, and esters, that one may call "physical" constituents. These are linked to molecules in the other compounds that generate the five tastes—sugar, acid, salt, bitter, and "umami"—to characterize a wine's or food's individual taste profile.

Umami is the most recently discovered taste. It was identified in 1908 by research scientist Dr. Kikunae Ikeda of Tokyo Imperial University.

The mango taste in your yogurt comes partially from the mango's sweet flavor, but mostly from its various aromatic molecules. Once in your mouth, after helping to provide the physical sensation of mango yogurt, these molecules become gaseous and waft into your nose, where their aromas complete the identification process.

## "MOTHER" FLAVOR

It has been shown that flavors pass from the mother to the baby via the amniotic fluid after the eleventh week of pregnancy. So, well before our birth, we have already become accustomed to several aromas and flavors coming from our mother's diet.

## AROMAS: THE GOLD RUSH!

Today, just as with spices in the days of Marco Polo, the biochemical molecules in flavorings can be worth more than their weight in gold. Some compounds are priced in the stratosphere!

**"Perfume is the most intense form of memory."**

JEAN-PAUL GUERLAIN

The flavor industry, in particular for processed foods, is an $8-billion-a-year industry. Some synthetic molecules have an even stronger taste than their natural equivalents. One example is ethylvanillin, whose taste is three or four times as strong as vanilla.

Two-thirds of the American diet is made up of processed foods, most of which contain flavor additives.

In Cincinnati, Givaudan (www.givaudan.com), the world's largest manufacturer of flavors and fragrances, makes over 6,000 versions of "strawberry," as well as 4,000 types of "orange," 3,000 flavors that taste like "chicken," and several thousand "butter" fragrances.

These flavors are created from molecules present in each of these foods, as well as in many other ingredients sharing their aromatic identity.

This proves the solidity of my thesis on food harmony and molecular sommellerie, which results in whole new families of complementary foods and wines.

## THE ADDED VALUE OF SMELLS AT THE TABLE

Without a doubt, smell gives a dish its complexity and provides its intrinsic qualities. The flavor of foods and beverages, including wines, depends on one's olfactory perception, which is the result of aromatic molecules released in the palate and transmitted via the nasal passages to the olfactory mucous membrane.

Food biochemists agree that aroma is *the* cornerstone of a food's overall flavor, from which the alchemy of the ingredients and their interactions provide people with the desire to eat or forego a given dish or ingredient. And the same holds true for wine.

Smell and taste serve as both negative reinforcements (disgust, that full feeling) and positive ones (desire, hunger). The choices we make as to which foods and wines to consume and which ones to push away are dictated by their aromas. Whether in the grocery store or the restaurant, fragrance determines our choices.

Among insects, which only eat to survive and to reproduce, eating is the result of sensorial stimuli that are mostly olfactory.

When we savor a wine, a dish, or a certain ingredient, the taste process passes through several steps. Simply put, we taste in two phases. *Primo,* via the nose and its perception of aromas. *Secundo,* via the mouth, as a result of the aromas' arrival through the nasal passages.

## OLFACTORY RECEPTORS IN CELLS—SPERM CELLS!

"It has been found that olfactory receptors aren't restricted to the olfactory neurons. This same type of receptor is found in several organs of the human body, such as the liver, and in the sperm-generating cells, where they may play an important role in guiding the sperm towards the egg. The sperm may be redirected during the course of their voyage to the egg by specially selected odiferous molecules, in effect opening new perspectives for local contraception." (*Revue des oenologues*, 2006)

To identify the general flavor of a food or wine, and thus form an impression of it, our brain combines these complex aromatic sensations with the collection of tastes, including the five fundamental tastes, as well as their interactions and aromas both in the nose and in the mouth. The syntax of these sensations plays a final role in combining it all and giving it meaning as a flavor.

The sense of smell is undeniably the most sensitive and subtlest of the senses. Humans are far from being the species with the most powerful sense of smell. Nonetheless, the human nose has proven to be, at least to date, more sensitive than most existing physico-chemical sensors in detecting aromatic molecules.

## UNIQUE AROMATIC MOLECULES!

In the molecular world of volatile compounds, it is remarkable to note that we have never found two different molecules with identical fragrances. Of course, molecules with similar structures may possess comparable fragrances, but their fragrances are still distinguishable. On the other hand, we know of several cases in which infinitesimal changes in the molecular structure can produce totally different fragrances. Inversely, structures that are quite different may produce similar fragrances.

## AN ERRONEOUS MAP OF TASTE

The scientific study of the chemical senses of smell and taste has long been neglected compared to that of the senses of sight and hearing. Fortunately, in the last fifteen years or so, studying the sense of smell has become considerably more popular. Linda Buck and Richard Axel's 1991 paper on the gene family corresponding to olfactory receptors and the organization of the olfactory system has opened the road to numerous scientific studies.

The 2004 Nobel Prize in Physiology or Medicine rewarded the contributions of Buck and Axel and inspired numerous researchers to continue their work on understanding the scientific mechanisms of smell and taste.

It is thus only recently that food science researchers have understood that the mapping of tastes perceived by the tongue, as established in the nineteenth century, is partially incorrect.

The idea that the tongue's taste buds were divided into four groups (only four basic tastes were known at the time), distributed in four strategic locations on the tongue and the wall of the mouth, is largely incorrect.

We know today that the different regions of the tongue can recognize several tastes at once. Sweetness is not recognized only at the tip of the tongue, nor is bitterness recognized only by the taste buds located at the back of the tongue and the palate.

When it comes to bitterness, we can no longer refer to it in the singular, but must instead use the plural, for there are at least five or six types of bitter.

## AROMATIC COMPOUNDS WITH A HIGH MOLECULAR DENSITY

The higher concentration of alcohol (ethanol) in fortified wines such as Madeira, port, and sherry has an impact on the volatile compounds that contribute to those wines' smell and taste.

Several volatile compounds become more soluble as a wine's alcohol level increases. Furthermore, some aromatic compounds become more difficult to smell when a certain alcohol level is attained.

On the other hand, a high level of alcohol allows aromatic compounds with a high molecular density, which are slower to become volatile in low-alcohol wines, to become more noticeable, and thus to contribute more to the taste of these high-alcohol wines.

In a low-alcohol wine, the first perceptible aromas at the wine's surface are the aromatic compounds with low molecular densities. In a wine with a higher degree of alcohol, such as fortified wines, few compounds of low molecular density are found at the wine's surface.

When savoring wines with "normal" alcoholic content, those between 10% and 14%, the aromas are released in successive stages. First come the compounds of low molecular density, and then later, depending partly on the impact of the oxygen in the air on the wine, come the high-density compounds. In higher-alcohol wines, those between 15.5% and 20%, the low-density aromatic compounds are rapidly inhibited, leaving room for the high-density compounds.

## DRINKING OR SAVORING?

The molecular density of volatile compounds largely explains why wines are so often perceived differently from one person to another, especially between those who linger and those who don't take the time to dissect a wine. There's a world of difference between "drinking" and "savoring."

If you put a sprig of rosemary under your nose, you will quickly recognize its general odor and you will note that it's rosemary. But you can also take a few moments to dissect the various compounds of its general fragrance and thus discover that rosemary contains wood, floral, and spice tones as well as notes of camphor and eucalyptus. That's the difference between "smelling" and "experiencing," between "drinking" and "savoring."

## SAVORING BY OLFACTORY "IMAGES"

In 2006, the *Revue des oenologues* announced the finding that the recognition of smells is similar to the recognition of forms. It's as though olfactory "images" are projected into the middle of the olfactory bulb.

The high number of olfactory receptor types, the combinatorial character of the information, and this "form recognition" characterization explain our ability to distinguish the odors of an incredibly high number of different aromatic molecules.

## MORE THAN 40 MILLION FRAGRANT MOLECULES!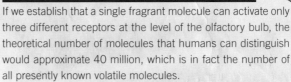

If we establish that a single fragrant molecule can activate only three different receptors at the level of the olfactory bulb, the theoretical number of molecules that humans can distinguish would approximate 40 million, which is in fact the number of all presently known volatile molecules.

## SCENT: SINGULAR OR PLURAL?

Some people incorrectly believe that each food has a single scent, that it is composed of a single aromatic molecule that provides its specific fragrance. On the contrary, the aroma of a single ingredient comprises an ensemble of volatile molecules that, when mixed together, provide its signature olfactory sensation.

Sometimes certain aromatic compounds dominate the mixture and consequently provide the principle aroma; examples include eugenol for cloves, cinnamic aldehyde for cinnamon, anethole for star anise and fennel, and thymol for thyme. But these examples are relatively rare.

In almost all cases, it's the mixture of aromatic compounds that provides the characteristic smell. For example, coriander seeds (including, among others, the following major volatile compounds: pinene, 3,7-dimethylocta-2,6-dienal [citral], linalool, and camphor) are both floral and lemony, and brushed with a touch of pine.

So the next time you smell an herb or a spice, take the time to really smell it, and even to experience it. You will discover an unexpected aromatic complexity. The same holds true for wines.

## THE NOBILITY OF BITTER TASTES

Of all the tastes, we recognize bitter ones most easily. For many professional tasters and gastronomists, these are the noblest tastes, and so they have the honor of closing this initial introduction to the universe of aromas and flavors.

We inherited the capability to effectively distinguish bitter tastes from our Paleolithic ancestors, who, by being able to recognize such tastes, were able to avoid food poisoning (plants with deadly compounds are usually very bitter).

## STRESS DISTURBS TASTE

Scientific research on food tasting shows that experimental subjects who are quite nervous have more trouble distinguishing bitter and salty tastes than do other tasters.

On average, humans are 1,250 times more sensitive to quinine, a very bitter substance, than to sweet-tasting sucrose. Funny, it would be easy to imagine that given our strong desire for sugar (mother's milk tastes of vanilla), we would be more sensitive to sugary tastes.

In fact, no! We are most sensitive to bitter tastes. I am using the plural because there are at least five or six different bitter tastes, and perhaps even more.

There is hope on the horizon for those who are repelled by bitter tastes, because tastes are not permanently determined by our genes. On the contrary, the tasting function is very flexible. Tastes change according to one's age, health, training, knowledge, and even state of mind.

So have confidence in your taste buds, and don't forget that the volatile molecules are talking to you!

OAK

SCHR

ROSEMARY

EF

MAPLE SYRU

MINT

FINO & MANZANILLA SHERRIES

AC

SOTOLON

AN GINGER

QUEBEC CHEESE

INNAMON

YCHEE FINO AND O

EWURZTRAMINER

STRAWBERRY

INEA

EN ANISE

EHYDES

CHERVIL

H FENNEL

WALNUTS

COUMARIN

PHENOLIC
ALDEHYDE

CINNAMON

CINNAMIC
ACID

OLOROSO
SHERRY

BENZOIC
ACID

PARSLEY
ROOT

GENEPY

ANISE-FLAVORED/MINT/
SAUVIGNON BLANC

AGASTACHE
ASAFOETIDA
BEE BALM
BASIL
DILL
CARAWAY
CELERY
CHERVIL
FRESH CORIANDER
FRESH FENNEL
LEMON BALM
LOVAGE
MINT
PARSNIPS
STAR ANISE (CHINESE BADIAN)
YELLOW BEETS
YELLOW CARROTS
OTHER VEGETABLES
ROOTS,
AS WELL AS PARSLEY,
LICORICE AND SHISO,
ALL CONTAINING
SAPID MOLECULES
WITH AN ANISE-LIKE TASTE.

ALBARIÑO (RÍAS BAIXAS/SPAIN)
CHARDONNAY (NON-OAKED/COOL CLIMATE, CHABLIS,
NEW ZEALAND)
CHENIN BLANC (LOIRE VALLEY AND SOUTH AFRICA)
CORTESE (GAVI/ITALY)
FURMINT (TOKAJI/HUNGARY)
GARGANEGA (SOAVE/VENETO)

# MINT AND SAUVIGNON BLANC

## AN OPEN DOOR INTO THE WORLD OF ANISE-FLAVORED FOODS AND WINES

> "The art of the good researcher is above all to ask the right questions. Einstein was a real master."
>
> HUBERT REEVES

### MINT: THE FIRST AROMATIC TRACK

In the course of my research, I tried to identify those molecules responsible for the flavors of given foods, especially the harmonic bridge ingredients. I noticed that foods could be divided into large aromatic families so as to provide greater precision when looking for harmonious wine and food pairings or inventing new culinary dishes.

We can note, for instance, a particular correspondence between mint and Sauvignon Blanc, as well as among dishes dominated by foods rich in volatile anise-like compounds, such as mint.

### BIRTH OF A DISCIPLINE

After having delved into the aromatic world of vin jaune (yellow wine) and curry (see the "Sotolon" chapter), the harmonious pairing of mint and Sauvignon Blanc was the first aromatic track that I explored when I began my work on molecular harmony in 2006.

Several years ago, I noticed that when mint is the dominant taste in a dish such as Middle Eastern tabbouleh (a refreshing bulgur salad with mint and fresh parsley) or a goat cheese sandwich with fresh mint and slices of cucumber and green apple, those dishes pair perfectly with wines, such as Sauvignon Blanc, that are often typified by an anise-like aroma belonging to the mint world.

So I started out trying to establish a molecular understanding of mint and Sauvignon Blanc in order to determine the reason for their mutual attraction. Later on, to my great surprise, I was able to establish (first on paper, later in the kitchen) the multiple aromatic links among mint and other herbs and vegetables sharing the anise-like taste of mint and Sauvignon Blanc. I started with green basil, fresh fennel, celery, chervil, parsley, and certain root vegetables such as carrots. When I tasted them individually, these foods always seemed to have a touch of anise. But to establish the links uniting them, I had to make a more "scientific" connection.

In the course of my work, while creating tables organized by volatile compounds, I came to understand that all these ingredients were interconnected by an assembly of similar aromatic molecules possessing a strong attraction for one another.

It took only a single step to proceed from theory to practice, which I quickly did. I thus succeeded in creating vibrant pairings between these "new" anise-flavored ingredients and Sauvignon Blanc or other wines with a similar profile. These pairings always proved to be just as resounding as the original one between mint and Sauvignon Blanc.

## HARMONIC PROOF...

I had right before my eyes (and on my taste buds!) both the theoretical and practical confirmation that my pairing theory made sense. Since then, I've taken it upon myself to survey the greatest possible number of wines and foods with the goal of uncovering new harmonic families. This knowledge offers a new road to creative cuisine, both for professional chefs and for inexperienced cooks. This process is well under way and continues to inspire me. I hope that this book, which offers a sampling of my research, will be the first in a long line to come.

## HERBS AND VEGETABLES TASTING OF ANISE

Herbs and vegetables with an anise-like taste usually come from one of three families: Apiaceae (Umbelliferae), which includes chervil and fennel; Asteraceae, which includes tarragon; and Lamiaceae, which includes basil and peppermint.

These basic ingredients contain anise-flavored compounds such as anethole (green anise, badian [Chinese star anise], green basil, celery, chervil, fresh fennel), R-carvone (mint), S-carvone (caraway), estragole (anise, basil, tarragon, fresh fennel, apple), eugenol (Thai basil, green basil, cloves), apigenin (parsley), and menthol (basil, fresh coriander, fresh fennel, mint, and root vegetables).

Note this interesting but relatively rare fact: R-carvone and S-carvone have different smells, but their molecules have the identical chemical composition with slightly differing structures.

## RHIZOMES AND ANISE-LIKE TASTES

Rhizomes (daikon, black radish, galangal, turmeric, ginger, chicory root, yam, parsley root, Jerusalem artichoke) also taste like anise, which links them to licorice and other ingredients in the anise flavor family.

## AN AROMATIC AFFINITY ON THE PLATE?

To confirm the true aromatic affinity of ingredients that had not already been classified in the anise category, I had to try out different wines at the dinner table.

These tests were conclusive. I would even say they were astounding and inspiring. Several different ideas for recipes involving these ingredients burst into my "psychological palate" (taste memory). This set the stage for a harmonious combination of ingredients on the plate, which in turn led to more vibrant recipes and an aromatic osmosis with the selected wine. This process culminated in wine and food pairings that were just as harmonious and yet even more precise than ever before.

Let's take, for example, my idea of redefining the mythical "Gargouillou of young vegetables," a signature dish of Michel Bras, the famous chef of the restaurant Laguiole in Aubrac, France. The idea was to recompose this jumble of vegetables and herbs to include only ingredients that taste of anise—in other words, foods from the Apiaceae, Asteraceae, and Lamiaceae families.

The result was a made-to-order salad (what a salad!) in perfect harmony with a white wine with the same anise-like tones—for example, a young, dry, non-oaked white wine based on Sauvignon Blanc or Verdejo. The dish is also compatible with wines based on Albariño, Greco di Tufo, Vermentino, Pinot Blanc, Furmint, Chenin Blanc, and Romorantin.

## SOMMELIER-COOK'S HINT

**Gargouillou of young vegetables "en mode anisé" (anise-flavored).** Assemble this fresh vegetable extravaganza with a seasonal selection of: bulbs of fresh or braised fennel, fennel fronds, celery (either in salt water or preserved), puréed Jerusalem artichokes, sliced Jerusalem artichokes steamed in licorice water, yellow beets, curried carrots, parsnips, celeriac, crosnes, salsify (vegetable oyster), parsley root, cucumbers, and green and yellow sweet peppers. Play with different fragrances by adding cubes of gelatin, dill, basil, chervil, fresh coriander, mint, and parsley, and don't forget a touch of olive oil scented with basil, fresh parsley, ginger, galangal, or turmeric. There are many possible variations on this kitchen garden recipe.

## SALMON CONFIT WITH PARSLEY, FENNEL, MINT, AND ROOT VEGETABLES

This idea is quite simple: prepare a salmon confit doused with parsley oil and accompanied by a salad consisting of raw fennel bulb—slightly blanched, very thinly sliced with a mandoline, and subtly enhanced by oil perfumed with fresh

# 1.
## COMPLEMENTARY
## FOODS
### ANISE-FLAVORED/MINT/SAUVIGNON BLANC

ANISE-FLAVORED/MINT/
SAUVIGNON BLANC

DANDELION

SAVOY CABBAGE
CULTIVATED

(STRONG ANISE TASTE)

CHICORY
(HELIOTROPE)

CUMIN
(FALSE ANISE FROM
THE MIDDLE EAST)

SEA
FENNEL

ENDIVE

FRESH PARSLEY

IRONWORT
(HYSSOP-LEAVED)

FRESH
CORIANDER

PARSLEY
ROOT

JERUSALEM
ARTICHOKE

FRESH FENNEL

YELLOW
BEETS

CELERY

MINT

ESCAROLE

LEMONGRASS

ABSINTHE

AGASTACHE
(MEXICAN TEA)

PARSNIPS

TARRAGON

OREGANO
AND SHISO
(PERILLA)

DILL

HYSSOP

CURLY
LETTUCE

BASIL

TREVISO
(RADICCHIO)

CHERVIL
(MUSKY CHERVIL)

STAR ANISE

LEMON
BALM

CELERIA

GARDEN
ANGELICA

ROOT
VEGETABLES
(RHIZOMES)

CARAWAY

YELLOW CARROT

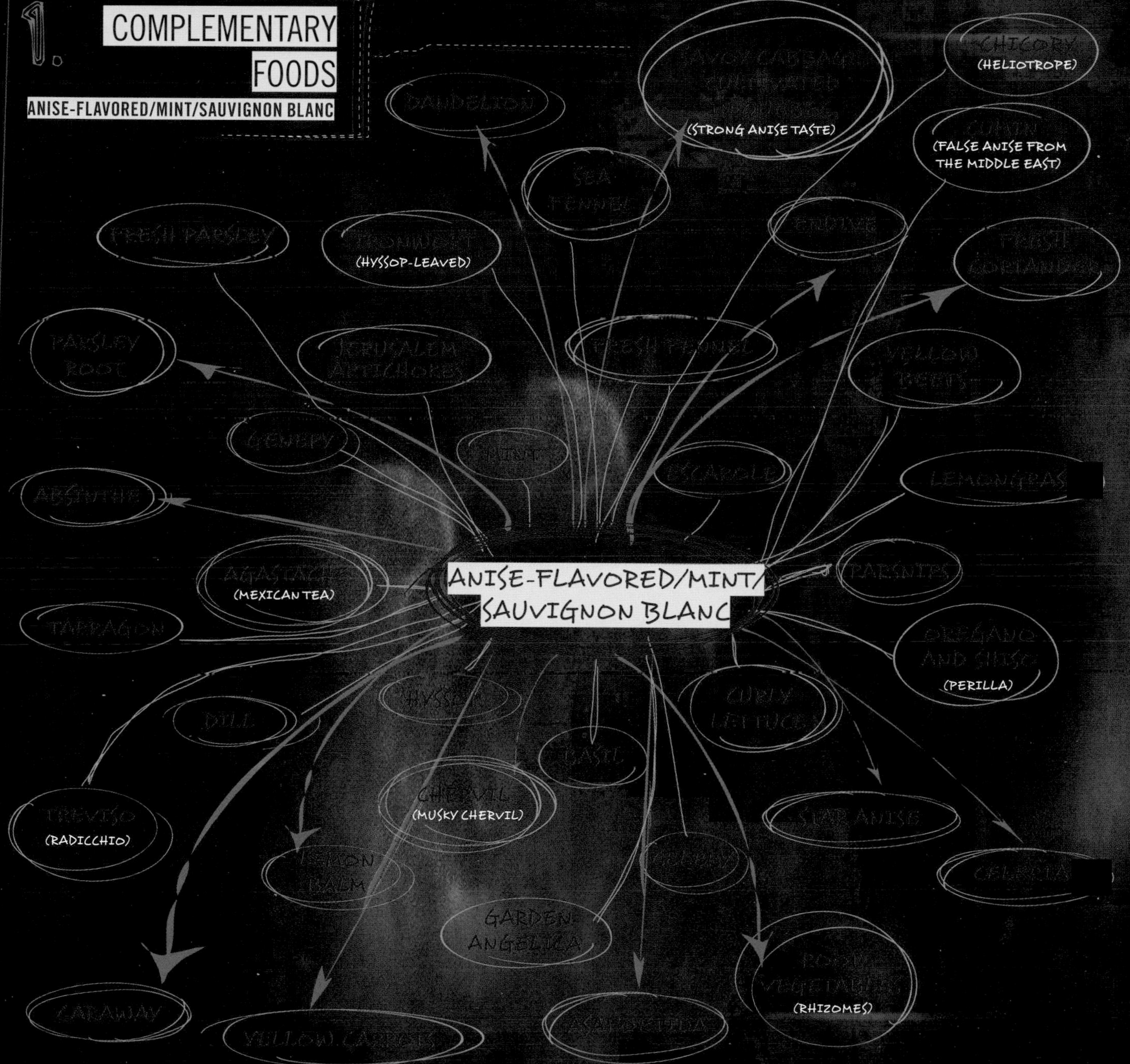

mint—accompanied by a purée of root vegetables such as parsnips or Jerusalem artichokes.

Pair the meal with a wine such as Sauvignon Blanc, and you'll easily have an almost perfect union of elements. No more false notes, as so often happens when the accompanying vegetables don't match the wine selected for the fish or meat.

All too often, wine is chosen to go with the main element of a dish (meat or fish, for example) without consideration of the vegetables, sauce, and other meal components. A fuller understanding of food flavors and wines makes it easier than ever to take into account all components of a given dish. Doing so will increase the meal's harmony and the success of the wine pairing.

### SOMMELIER-COOK'S HINT

**An anise-flavored sandwich that's quite cool.** Make a goat cheese sandwich with thin, crunchy slices of green or red apple, accompanied by a julienne of fresh fennel (or slices of yellow beets) and cucumber, accompanied by fresh mint and mayonnaise with caraway seeds or wasabi and perhaps a slice of smoked trout. Serve with a fine glass of Sauvignon Blanc or Verdejo. This great pairing is simply delicious!

This recipe is easy as pie, really refreshing, and representative of recipes based on the anise family that are cool-tasting in more ways than one. It proves beyond a doubt that the results of research on food harmony and molecular sommellerie are adaptable to any circumstances, both for day-to-day meals and for gourmet repasts. It works for low-cost recipes and wines, and for those expensive blowouts as well. Regardless of your budget or your level of skill in the kitchen, you'll always achieve harmony.

# VOLATILE COMPOUNDS AND AROMAS
### ANISE-FLAVORED

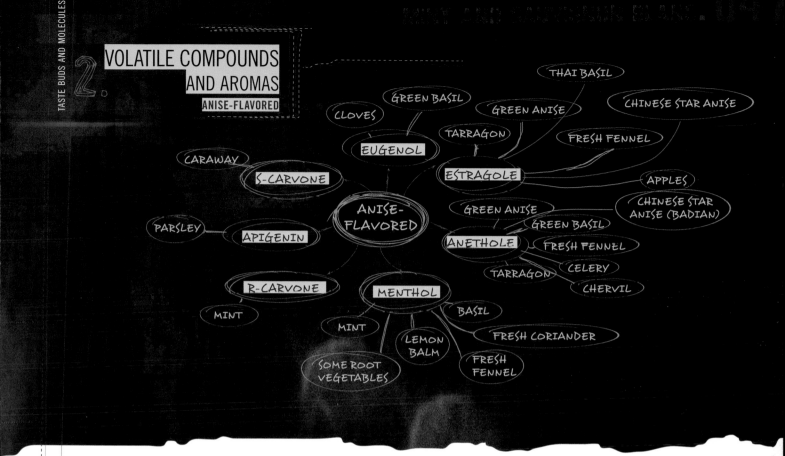

THAI BASIL

GREEN BASIL

CHINESE STAR ANISE

CLOVES

GREEN ANISE

TARRAGON

FRESH FENNEL

EUGENOL

ESTRAGOLE

APPLES

CARAWAY

S-CARVONE

CHINESE STAR ANISE (BADIAN)

ANISE-FLAVORED

GREEN ANISE

GREEN BASIL

PARSLEY

APIGENIN

ANETHOLE

FRESH FENNEL

CELERY

TARRAGON

CHERVIL

R-CARVONE

MENTHOL

BASIL

MINT

MINT

FRESH CORIANDER

LEMON BALM

FRESH FENNEL

SOME ROOT VEGETABLES

## ANISE-FLAVORED MOLECULES

As I mentioned before, volatile anise-flavored compounds are chiefly found at various concentration levels in vegetables and herbs of the following families: Apiaceae (umbellifers), Asteraceae, and Lamiaceae. Figure 2 shows some of the major volatile compounds in these anise-flavored foods.

The dominant molecules of anise-type aromas are anethole and estragole, which are found chiefly in Chinese star anise (badian), green basil, fresh fennel, celery, chervil, and tarragon, and to a lesser extent in fresh apples.

Foods rich in anethole include dill, anise, Chinese star anise, green basil, caraway (meadow cumin), tarragon, and fresh fennel. They have in common several other molecules that give them their distinctive character.

Yellow beets also have the taste of anise, but that isn't really the case for red beets, which are fruitier and earthier tasting due to the presence of geosmin, a volatile molecule with a strong odor of wet earth/rotting wood, an aroma considered a fault when found in wine.

## "COLD-TASTING" FOODS

Mint is one of a group of "cold-tasting" foods. I chose this term because of various aromatic compounds such as menthol, or estragole in apples, that provide a cooling sensation in the mouth.

Menthol occurs in all types of mint. A small quantity sets off the mouth's thermal cold receptors. A large quantity has a burning effect, the way capsaicin does in hot peppers. Just like apples, green peppers, and cucumbers, mint contains other molecules that, like menthol, stimulate those receptors

which are also activated by temperatures between 46°F and 82°F (8°C and 28°C). These ingredients stimulate cold, providing the sensation of coolness in the mouth, especially when eaten raw.

Since mint is part of the anise-like family, we can add to these four cold-tasting ingredients (see the chapter "A Taste of Cold") several other refreshing, anise-like foods such as celery, parsley, fennel, coriander, parsnips, and parsley root, as well as ginger, green apple, lemon balm, lemon verbena, and lemongrass (which is close to lemon balm and lemon verbena).

These ingredients belong to other families of aromatic compounds, but they create a perceptible sensation of freshness in the mouth quite like that of the anise family.

This sensation of cold should be considered when selecting a wine pairing. Depending on the meal's other elements, their presence can act as a buffer, for example, calming the spices' fires. We also must consider the serving temperature of the food. The best choice is therefore a wine rich in alcohol or a wine served at a somewhat higher temperature, because the cold-tasting foods have already created a coolness in the mouth.

### WHITE WINES WITH AN ANISE-LIKE PROFILE

The dominant touch of anise, often provided by volatile molecules tasting of mint, basil, or fennel, calls for serving certain white wines known for their refreshing acidity and electrifying minerality. Such wines pair well with dishes containing lots of anise-flavored foods, some of which also contain cold-tasting elements, as previously described. These foods include dill, green basil, chervil, celery, fresh coriander, fresh fennel, mint, and parsley, as well as carrots, yellow beets, parsnips, and other root vegetables.

The ingredients in this family are in complete harmony with a young, dry, preferably non-oaked white wine presenting anise-like notes (see Figure 3).

### RECIPE FOR ANISE-FLAVORED SPRING ROLLS

To put into practice the theoretical results of my research on the anise family, I cooked one of my favorite sunny-day foods: shrimp spring rolls. Anyone can prepare this at home; it's as easy as knowing how to roll.

To achieve an almost perfect pairing with a white wine like Sauvignon Blanc—I chose the spectacular 2000 vintage of Silex Blanc Fumé de Pouilly, produced by the late Didier Dagueneau, which was refreshing, ethereal, and extroverted, with a vibrant anise-like subtlety and a unique satiny structure—I added to the shrimp rolls (having pre-sautéed the shrimp in a frying pan) a thinly sliced julienne of daikon, slivers of celery hearts and leaves, thin strips of fennel root and

3. WHITE WINES
TASTING OF ANISE

ALBARIÑO
CHARDONNAY

CHENIN BLANC
CORTESE
FURMINT
GARGANEGA
GODELLO
GRECO DI TUFO
GRÜNER VELTLINER

PINOT BLANC
RIESLING
ROMORANTIN
SAUVIGNON BLANC

VERDEJO
VERMENTINO

fronds, and finely chopped baby sweet peas and their pods, and finished it off with cucumber, green and red pepper, and, of course, several fresh mint leaves.

These anise-family ingredients and cold-tasting ingredients—cucumber, sweet peppers, and mint—linked beautifully with this great, remarkably long-lasting wine. Finally, to seal the pact, especially with a Sauvignon Blanc, I grated several very thin blanched lime zests, which I added to the stuffing just prior to rolling the mixture. The lime's essential oils perfumed the contents of the rice sheets and constituted one more bridge to the wine with its notes of citrus (terpenes and linalool).

## PASTIS

Pastis is a liqueur traditionally made from licorice root and star anise. The fruits of star anise are much richer in anethole, a compound with an anise-like taste, than is fennel. So, foods dominated by pastis, such as a shrimp vol-au-vent in pastis, also harmonize very well with these wines.

## RED WINES WITH AN ANISE-LIKE PROFILE

Anise-like ingredients are not only found in dishes created to be paired with white wines; we must also consider pairing them with red wines. When a recipe's anise-flavored elements (such as fennel, star anise, mint, pastis, parsnips, black licorice extract, and Jerusalem artichokes and other rhizomes) require a red wine, go for wines based on Syrah or Shiraz and Grenache/Syrah/Mourvèdre (GSM) blends. Such wines are usually rich in anise-like scents, such as those of licorice, anise, mint, and tarragon.

## AN ANISE-FLAVORED MEAL PAIRED WITH RED WINE

Braise a lamb shank with tomatoes, pastis, and fennel and serve it with a Syrah from the Rhone Valley or from South Africa. Both wines express the same anise tonality found in this lamb dish dominated by pastis and fennel, both of which are rich in anethole.

Ingredients in the anise taste family are in perfect harmony with red wines with anise-like notes, such as those listed in Figure 4.

## BLACK LICORICE EXTRACT

We must add black licorice extract to the list of anise-like tastes that possess a molecular power of attraction with the above red wines. Anyone who has followed my wine reviews during the past several years will know that I often use black licorice extract when pairing food with Syrah, Grenache/Syrah/ Mourvèdre (GSM) blends, or Cabernet/Syrah blends.

Black licorice extract, when added to a sauce near the end of cooking, does an excellent job of softening the tannins in red wine.

Black licorice extract owes its softening power to one of its chemical components, gylcyrrhizic acid, which is largely responsible for its strong, persistent flavor. This molecule, helped along by other volatile licorice compounds such as maltol, a flavor enhancer widely used in the food industry (see the "Oak and Barrels" chapter), accelerates the wine's aging. In other words, licorice gives wines length and persistence in the mouth. Even moderately long wines take on greater length in the mouth when they are enjoyed with a dish containing black licorice extract.

## RED WINES
### TASTING OF ANISE

CABERNET-SHIRAZ (AUSTRALIA)
CABERNET-SYRAH (PROVENCE)
GARNACHA-TEMPRANILLO (SPAIN)
GRENACHE-SYRAH-MOURVÈDRE (LANGUEDOC-ROUSSILLON/RHONE/AUSTRALIA)

SHIRAZ (AUSTRALIA)
SYRAH (CROZES-HERMITAGE/LANGUEDOC/ITALY)

Black licorice extract qualitatively augments "small" red wines and literally amplifies "fine" red wines.

Black licorice extract—more precisely, its principal aromatic molecule, glycyrrhizic acid—is added to pastis during the production process in order to smooth it, to give it more presence, and to extend its anethole flavor. Without this acid, pastis would be tasteless and harsh, as are many of the cheap, poor-quality versions. This confirms black licorice extract's power of softening tannins and extending the flavor persistence of tannic red wines. Some dishes in which anise flavors predominate—for example, those based on tarragon—also have a softening and lengthening effect on red wines.

### LICORICE: AN EMULSIFIER

Licorice's glycyrrhizic acid is a saponin with very stable emulsifying properties. Saponin is also found in tubers, green peas, spinach, soybeans, quinoa, pulses, tea, corn, garlic, ginseng, tomatoes, and chestnuts (a particularly rich source), so it is possible to make stable mousses from these foods.

Black licorice extract is more complex than licorice twigs and is also richer in amino acids, which explains its marked presence in the mouth. Concentrated black licorice extract is even more complex, having had its principal aromatic compounds intensified by toasting. This concentrate is used, for example, to add color and aroma to certain dark beers, such as stout and porter, and to add aroma to some cigars.

Finally, black licorice extract contains estragole, a molecule with a cold taste, which makes it quite refreshing in the mouth. Estragole is also found in star anise, cinnamon, cloves, ginger, fennel seeds, apples, green basil, tarragon, sage, mustard, and bay leaves. One should consider all these ingredients when cooking with licorice, or when seeking similar foods to replace it, to increase a recipe's freshness and improve its wine pairings.

### ANISE-FLAVORED FOODS PAIRED WITH GREEN TEA

Wine is not the only beverage that can accompany anise-flavored foods; certain teas are also good candidates. If you are looking for adventures in tea drinking, savor a piece of chocolate flavored with star anise alongside a cup of green tea with a chervil-like aroma, such as Japanese Gyokuro Tamahomare. If your chocolate is fairly rich and unctuous, choose instead a green Chinese tea. They are richer in amino acids, and so are rounder and more mouth-filling, because the umami taste is present.

One of the best-known green Chinese teas, Bi Luo Chun, is extremely complex. Its dry leaves generate notes of chocolate. After steeping, they develop light notes of iodine. In your mouth, the texture is rich, unctuous, and marked by flavors of both vegetable (fiddleheads) and fruit (peach), with notes of dried fruit and dark chocolate.

You can also create this type of harmony with anise-flavored foods by making a mint tea with a base of green Chinese Gunpowder tea.

## NEW DIRECTIONS IN CREATIVE CUISINE

Once you understand this family of anise-flavored foods and wines, you can really let your imagination go. Experiment with combining ingredients to create your own recipes in harmony with the selected wines. Here are some suggestions:

**Carrot and apple salad with caraway seeds and paprika**

**Carrot cake with caraway seeds and celery-apple compote**
An entrée (more salty than sweet) or a dessert based on carrots flavored with caraway, a spice whose anise-like taste meshes well with this root vegetable, and apples with celery, which together create an acidic, invigorating taste.

**Munster à l'ajowan** The Alsatian dish known as *Munster au cumin* is actually prepared with caraway seeds (caraway is in the same family as cumin, but it is from the meadows of Eastern Europe and not from the Middle East). Since ajowan (sometimes also called ajwain) is related to caraway and has a similar taste, why not surprise your guests with a Munster cheese *à l'ajowan?*

**Dark chocolate with basil, fennel, or mint** The volatile compounds of basil, fennel, and mint, all members of the anise flavor family, go hand in hand, even when they are added to a dark chocolate ganache, with those of late-harvest Sauvignon Blanc wines such as the Chilean Errazuriz Late Harvest. Yes, it is possible to serve a white sweet wine with dark chocolate; basil serves as a pairing bridge to such wines.

**Spanish arbequina olive oil and recipes with an anise flavor**
The season's first olive oil, based on arbequina olives, tasting of green fruits and herbs with a touch of anise, is ideal for salads or served on bread with a glass of cool, anise-flavored white wine made from the Verdejo grape of Rueda, Spain. If you sprinkle all the anise-flavored ingredients with this oil, you'll increase their power of attraction for a Sauvignon Blanc or similar wine. When arbequina olive oils have aged for several months, they are better enjoyed with sweet dishes because to some extent they have lost their green fruit/fresh herb character.

## OTHER IDEAS FOR ANISE-FLAVORED DISHES

+ Ceviche of oysters with wasabi and coriander
+ Snails with mushrooms and creamed parsley
+ Filet of pink trout grilled in basil oil
+ Parsleyed mussels in their own juice
+ Pasta with lemon, asparagus, and fresh basil
+ Pasta with pesto
+ Pasta with smoked salmon and dill sauce
+ Grilled scallops and smoked eel with cream of celery
+ Parsleyed fried scallops in their own juice
+ Leeks braised in mint
+ Chicken breasts stuffed with Brie cheese and caraway
+ Shrimp and basil risotto
+ Sandwich with goat cheese, cucumbers, green apples, green pepper, and fresh mint
+ Trout with purée of celeriac
+ Shrimp vol-au-vent in Pernod

## SOME CANAPÉS WITH ANISE-LIKE FLAVORS

+ Shrimp puffs with creamed parsley
+ Brie puffs dusted with caraway
+ Canapés of smoked salmon and dill
+ Canapés of smoked trout on a bed of puréed celeriac
+ Parsnip fries with yogurt and cumin sauce
+ Mini brochettes of shrimp with basil
+ Mini brochettes of grilled scallops with celery salt

## OTHER FOODS WITH TOUCHES OF ANISE

**Lemongrass:**

Lemongrass, whose principal volatile compounds are citral, geraniol, and linalool, exudes aromas similar to lemon balm and lemon verbena. This explains the harmony between a Thai soup with lemongrass and a young, vivacious Riesling, or a young un-oaked Chardonnay tasting slightly of anise and a white fish steamed with dried lemon verbena. (Note that dried lemon verbena, unlike dried lemon balm, retains its aromas for a long time. These aromas are liberated by heat.) Another good choice here is a dry Muscat from Alsace.

**Lovage:**

This aromatic plant has a slight taste of anise that is reminiscent of caraway. It is also rich in thymol and in carvacrol, an essential oil of thyme (and also of ajowan, sage, basil, rosemary, and mint).

**Lemon balm and lemon verbena:**

In addition to their subtle anise profile, the lemony perfume of lemon verbena and lemon balm (whose aroma is more subtle than that of verbena) is characteristic of citral, a sweet-smelling molecule that can be found in lemon zest along with several other volatile compounds, such as limonene and citronellol. It is citral in particular that expresses the aromatic identity of lemon balm and lemon verbena. This is also true for the aroma of lemon zest, which reminds us of these two herbs even more than does the lemon itself.

Citral is also found in wines, especially young Muscats and Gewürztraminer. It is an intermediate compound that leads to linalool, another volatile compound with floral and citrus tones that are very present in these two grape varieties. Sweet Sauternes and Jurançons are also rich in linalool. Lemon balm and lemon verbena are often found in the complex aromas of some Rieslings and Chardonnays, which opens the door for yet more wine pairing possibilities with these two herbs.

**Shiso:**

Shiso, also known as perilla or Japanese basil, is a member of the mint family (Lamiaceae). It has a taste that is somewhere between fennel, mint, lemon balm, and licorice, with hints of cinnamon, and a slightly astringent taste. It also presents a subtle taste of curry. The variety known as purple or red shiso is less aromatic. Shiso has antiseptic properties, which explains its use in Japan for preserving foods such as fish. It also works against allergies by reducing the production of histamine and immunoglobulin E.

**SOMMELIER-COOK'S HINT**

**Shiso health juice** In Japan, in season, shiso health juice is popular. This beet-colored drink is made with shiso, vinegar, and honey. It can also be used in the kitchen to make an anise-flavored sauce.

FROM VIN JAUNE
TO SAUTERNES...
BY WAY OF
WALNUTS,
CURRY,
FENUGREEK,
MAPLE SYRUP,
BALSAMIC VINEGAR,
PRUNES,
VDN WINES,
WHITE AND TAWNY...
A WORLD OF FLAVORS
TO EXPLORE
IN THE AROMATIC
UNIVERSE
OF SOTOLON

COFFEE

NUTS

BROWN SUGAR

# SOTOLON

## THE MOLECULAR CHAIN LINKING VIN JAUNE, CURRY, MAPLE SYRUP, SAUTERNES, ETC.

> "Research is a long process
> of acquiring information."
>
> RICHARD BÉLIVEAU, PH.D.

I will begin this chapter dedicated to the molecule at the base of my research with a story that perfectly illustrates the path that led me to further my understanding of the molecular structure of foods and wines.

Over the course of many foggy hours spent decoding wine's volatile compounds, I became aware that the major aromatic note (curry and walnuts) of vin jaune (yellow wine) from Jura, France, originated in 4,5-dimethyl-3-hydroxy-2(5H)-furanone, better known as sotolon. This volatile molecule gives curry and walnuts, among other foods, their very particular taste.

Shortly thereafter, I discovered that this same aromatic molecule is dominant in fenugreek, more precisely in roasted fenugreek seeds, which are part of some curries and provide one of the most important aromatic tonalities to our Canadian maple syrup.

### CARAMEL/MAPLE = SOTOLON

With its powerful caramelized aroma of roasted fenugreek seeds, sotolon has long been used to simulate the odor of caramel and maple syrup. Today, sotolon is even available in chemical form.

I also discovered that sotolon is a factor in the aroma of flax (especially the oil) and *sous-voile* wines (wines raised under a cap of yeast in the barrel), such as some sherries.

Sotolon is also found in some red and white naturally sweet wines (raised in an oxidative milieu), sake, soy sauce, highly fermented beers (brown and black), beef boullion, and certain dried mushrooms (Lactarius); in the aromas of fine Havana cigars (infused leaves); and in lovage, molasses, black smoky or aged teas (wulong and pu-erh), mature rum, and maple syrup. As well, it is found in aged sweet white wines affected with noble rot—here we are not talking about *botrytis cinerea,* but simply about overripe grapes—in particular, syrupy white wines such as Sauternes and Tokaji Aszú.

### SOMMELIER-COOK'S HINT

**Matured, settled Sauternes, and a tarte tatin with curry topped with a slice of fried duck foie gras** My scientific discoveries explain this pairing, which I empirically created in 2002 at the closing gala of the charitable event Montreal Passion Vin (www.montrealpassionvin.ca). At that time, I paired a mature Chateau Rieussec 1979 Sauternes and a curried tarte tatin, topped just prior to being served with a slice of fried duck foie gras, presented as a dessert. I invented this dish expressly to pair it with this wine. The aromas of walnuts, curry, apple confit, and caramel of this Sauternes launched me on this harmonious wine pairing path in the aromatic world of sotolon.

CURRIED TARTE TATIN TOPPED WITH
A FRIED SCALLOP OF DUCK FOIE GRAS

After defining the type of products in which sotolon appears, it was easy to conclude that it is also characteristic in the aroma of tawny port, which is raised in oak for many years, and thus in contact with oxygen, just like the aromatic profile of some old ports and Madeiras. Over the course of their slow evolution, Madeira wines undergo heating favorable to aromatic molecules such as sotolon. In effect, heating them above 37°C (about 99°F) sets off important chemical transformations that can produce fragrances associated with the sotolon family.

## A LOOK IN THE REAR-VIEW MIRROR

More recently, in 2005, my readings on the history of wine reminded me that in ancient Rome, the most prized wines were old, sweet white wines that had taken on a taste of rancio (touched by sotolon…).

Because these wines were rare and expensive at the time, people added spices such as fenugreek and even salt water to sweet white wines to give them the desirable "old" taste of oxidized wine.

After several unsuccessful attempts with sweet white wines, I tried to revisit the past, all the better to foresee the future, by setting up in a workshop a vat inspired by the Romans, laced with sotolon. To do so, I had to find the exact dose of roasted fenugreek seeds to add to the mixture. A young, keen, dry white wine, dense and able to resist slight oxidation, proved to be the best choice for this experiment.

The perfect wine for this exercise was a Cuvée Marie 2004 Jurançon Sec, Charles Hours, France. With the required maceration time (more or less three days), it managed to acquire the aromatic profile of a mature, settled wine, in the aromatic sphere of sotolon (curry, walnuts, maple syrup, fenugreek), all the while maintaining its youthful spirit! This was a vibrant pairing exercise.

## FROM SOTOLON TO THE SPICE ROUTE

In February and March of 2007, at the request of the "alchemist of spices," chef Racha Bassoul of the now-defunct Montreal restaurant Anise, I had the great privilege of providing the inspiration for a menu with the theme "The Spice Route." This meant selecting the wines and guiding the choice of spices and the composition of dishes, always "for and with" the chosen wines.

I'll explain how I arrived at one of these creations, centered around sotolon and maple syrup, presented in the second act of this repast:

**Cuvée Sotolon —**
**Inspired by the Romans and conceived by François Chartier**

*Based on a young, dry Jurançon, in which roasted*
*fenugreek seeds have been macerated*
*to provide a more mature, spiced profile*
*in the image of classic Roman wine*

**and**

**Three Princesses and Three *Espumas***

*Princess scallops and their coral, and three types of*
espumas *(sake and salt water; roasted fenugreek seeds*
*macerated in wine; maple syrup), accompanied*
*by watercress and red shiso.*

In order to achieve the perfect pairing, I started with a dish composed of elements that share several volatile compounds, all in the sotolon family. There was iodine in the scallops, sake, maple syrup, roasted fenugreek seeds, and the salt water of the *espumas* (airy mousses resembling foam, produced by a siphon, without any fats).

Everything in the dish and the wine was inspired by the powerful odor of the roasted fenugreek seeds, dominated by soloton and, above all, the aromatic signature of maple syrup.

The pairing of the Cuvée Sotolon and the three Princess scallops was in fact the result of several years' research on linking ingredients and volatile compounds.

As so well expressed by Dr. Richard Béliveau, quoted at the beginning of this chapter, "Research is a long process of acquiring information."

# 1. COMPLEMENTARY FOODS
## SOTOLON

AGED AND SMOKED BLACK TEAS

BEEF BOULLION
BROWN SUGAR
CELERY SALT
COFFEE
COOKED CELERY
COTTON CANDY
CURRY
DATES
DRIED FIGS
DRIED MUSHROOMS
HAVANA TOBACCO

LOVAGE
MADEIRA
MAPLE SYRUP
MOLASSES
POWDERED MALT
PRUNES
REDUCED BALSAMIC VINEGAR
ROASTED FENUGREEK SEEDS
ROASTED NUTS
SOY SAUCE
TAWNY PORT

# 2. COMPLEMENTARY WINES AND BEVERAGES
## SOTOLON

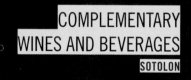

AGED BROWN RUM
HIGHLY FERMENTED BEER

MADEIRA (BUAL AND MALMSEY)
MATURE CHAMPAGNE
MATURE WHITE WINE
MONTILLA-MORILES
NATURALLY SWEET WINES

OXIDIZED WHITE WINE
PORT
SAKE
SAUTERNES
SHERRY
TOKAJI ASZÚ
VIN JAUNE
VINO SANTO

### NEW PATHS IN CREATIVE CUISINE

Now you understand a bit about the paths that inspired me to track down the molecular chain linking certain ingredients, wines, and beverages. So, what next? Well, as I did for the repast "The Spice Road," the key is to establish aromatic bridges between dishes dominated by a given ingredient and the wines or beverages that share compounds with them. Here are several examples:

**Curry-dominated recipes:** Pair with a vin jaune or a mature Sauternes.

**Dishes sweetened with maple syrup:** Pair with a mature Tokaji Aszú, a Vino Santo (sweet Italian), a Nigori sake, an oloroso sherry, or a Montilla-Moriles Pedro Ximénez solera.

**Salty dishes based on maple syrup:** Pair with a Vino Santo (the least sweet possible), a sake, an aged rum, or an oloroso sherry. If foie gras forms the heart of this dish, go with a mature Sauternes.

**Salmon brushed with soy sauce or balsamic vinegar:** Pair with a Montilla-Moriles amontillado or a manzanilla pasada sherry.

Now it's your turn to play and be daring in the kitchen! All you need to do is use the "molecular chain" of sotolon as a natural bridge between ingredients and between dishes and beverages. The door is wide open to creating new recipes containing ingredients that are rich in volatile compounds of the sotolon family, and to pairing them with wines and other beverages whose aromatic tonalities come from the same molecular sphere.

### SOME IDEAS FOR SOTOLON-BASED RECIPES:

+ Caramelized pineapple with a reduction of soy/sake/brown sugar, slivers of dark chocolate, and licorice powder
+ Cotton candy with maple syrup, flavored with roasted fenugreek seeds
+ Smoked meat glossed with soy/sake/brown sugar
+ Grilled beef with a soy sauce/maple syrup reduction
+ Caramel popcorn flavored with roasted fenugreek seeds
+ Curried tarte tatin topped (just before serving) with a slice of fried duck foie gras

### THE EMERGENCE OF SOTOLON-LINKED AROMAS IN WINE

Sotolon is most strongly present in *sous-voile* wines (fino sherry and vin jaune from Jura) as well as in wines that have undergone extensive oxidative aging. Oxygen is the element with the heaviest impact on the development of aromatic molecules in the sotolon family. Its volatile compounds, therefore, are present in varying degrees in wines rich in aldehydes, such as vins jaunes, vins de paille (straw wines), sherries, and old VDN wines based on black or white Grenache grapes. They are also present in sweet, syrupy wines, such as Sauternes, whose harvest was affected by noble rot *(botrytis cinerea)*, which includes the Tokaji Aszú (in particular the old, more oxidized examples or the more recent ones that are more than ten years old, such as the heavily affected 1996 vintage), old tawny ports, old white ports, old Bual and Malmsey Madeiras, and some old Champagnes that have matured on their lees.

### OLD DRY WHITE WINES AND SOTOLON

It has been shown that a dry wine (less than 5 grams of residual sugar per liter) stored under conditions of intensive oxidation will have a large concentration of sotolon. The best example of this is the vin jaune from Jura. In fact, every white wine stored in the presence of oxygen will develop volatile compounds of the sotolon family.

### THE AROMATIC IMPACT OF CYCLOTENE

Roasted fenugreek seeds contain molecules of the cyclotene family, which are found in plants as well as in maple syrup.

Cyclotene, whose strong odor is reminiscent of maple syrup and to a lesser extent licorice, also resembles, at least aromatically, furanone (a compound even more powerful than sotolon, which has a smell of carmelized maple syrup and is dominant in soy sauce) and maltol (burnt sugar).

Thus, all grilled or roasted foods containing sugar (grilled almonds, cocoa, coffee, roasted chicory roots, maple syrup) are filled with aromas associated with cyclotene, just as with sotolon.

Cyclotene may also be responsible for the mineral taste of flint, similar to silex, that is found in certain white wines. A very mineral vin jaune (smelling of flint), endowed with a pronounced "yellow" taste, stamped with sotolon and expressed in notes of curry, fenugreek, and walnuts, can be harmoniously paired with a fish containing notes of iodine and flavored with fenugreek or maple syrup (or soy sauce, sake, flax oil, etc.).

## SOMMELIER-COOK'S HINTS

**Manzanilla with fenugreek** Why not add some roasted fenugreek seeds to a young, heavily iodized manzanilla sherry to generate a developed iodized wine such as in the days of ancient Rome, but one that is sweeping and vigorous like a manzanilla pasada?

**elBulli: The transformation of the oyster** Have fun with elBulli's emblematic oyster meringue, a signature dish created in 1995, which consists of foamed sea water deposited on an oyster, accompanied by roasted fenugreek seeds. Pair this dish with a manzanilla pasada sherry, such as the superb La Bota de Manzanilla Pasada 10.

**Recipe for elBulli's oyster meringue**
This oyster meringue is served in a Chinese porcelain spoon, garnished with a thin julienne of smoked pork shoulder and shallots, and coated with a foam of sea water. The foam is obtained by recovering the water expelled when the oysters are opened and then adding gelatin. Let the mixture sit for twelve hours in a siphon. At serving time, add a light layer of airy mousse from the siphon to the oyster. Add a second layer of foam, this time flavored with roasted fenugreek seeds. (For the complete recipe, see the book *elBulli 1994-1997: A Period That Marked the Future of Our Cuisine.*)

## RED AND WHITE VDN WINES
## RAISED IN AN OXYDATIVE MILIEU

Finally, French red and white VDN wines, made from white, gray, or black Grenache grapes (and to a lesser extent from Maccabéo and Tourbat), so not from Muscat—some examples being Banyuls, Maury, and Rivesaltes—are vinified in an oxidative milieu. Such wines mature outdoors for a long period in oak, and often in the bottle as well. This slow and lengthy maturation process in contact with oxygen helps develop volatile compounds of the sotolon family.

## RANCIO = SOTOLON

Sotolon is responsible for the famous rancio, often associated with the aroma of rancid walnuts, that is so characteristic of mature VDN wines. The more discernible the rancio characteristic when savoring these wines, the higher the level of sotolon.

Sotolon concentrations are much higher in VDNs raised in the presence of oxygen than in those raised in the absence of oxygen (vintage or rimage). A red, fifteen-year-old vintage VDN, preserved without oxygen—in other words, quickly bottled—has a very low level of sotolon in comparison to a VDN of the same age that was conserved in a "moderately" oxidative milieu (partially oaked).

A seven-year-old rancio VDN preserved in a heavily oxidative milieu (perhaps in outdoor oak barrels or in the bottle) has a much higher sotolon concentration than the two preceding types of VDN. Oxygen is fundamental to sotolon's development. Sotolon concentration also greatly increases in white VDNs raised in the presence of oxygen. In fact, white VDNs raised in an oxidative milieu present the highest levels of sotolon. At the same age, a white rancio VDN contains twice as much sotolon as a red rancio.

## DRIED FIGS AND VDN WINES

The more that aged VDNs are characterized by the odor of dried figs (see the "Fino and Oloroso" chapter), the richer they are in sotolon. So we should pair old non-Muscat VDNs raised in an oxidative milieu with dishes dominated by dried figs, prunes, grilled almonds, honey, cinnamon, dried fruit, and vanilla, as well as with any ingredient or beverage influenced by the presence of sotolon.

CHEESE WITH ROSEMARY AND FINO
AND MANZANILLA SHERRY

LUSTAU

LACTONES

SOLERONE

DRIED
FIGS

LINALOOL
AND OTHER
TERPENES

# FINO AND OLOROSO

## A VEIL OF AROMAS WITH SHERRY WINES

"The affirmation of individual taste is a quest for freedom."

JACQUES PUISAIS, OENOLOGIST, FOUNDING VICE-PRESIDENT OF L'INSTITUT DU GOÛT (THE TASTE INSTITUTE),
HONORARY VICE-PRESIDENT OF THE INTERNATIONAL ASSOCIATION OF OENOLOGISTS, AND AUTHOR

Hot, bewitching Andalusia, situated on the southern border of Spain, includes two appellation zones, which furnish wines of many personalities. One of these is Jerez, the internationally known appellation whose wines, also called xérès or sherry, are found in the Cadiz region. The other appellation is Montilla-Moriles, located south of Cordoba, more than a hundred kilometers (some sixty miles) east of Jerez.

Both of these areas produce wines whose alcohol levels vary from 15% to 20% and which partially mature in ingenious collections of barrels known as *solera*. This system consists of a clever assembly of young wines and much older ones, creating wines whose character is quite varied and interesting.

### A QUESTION OF STYLE

Some wines—in this case lightly fortified (15%) fino and manzanilla, a specialty coming from Sanlúcar de Barrameda in the region west of Jerez—produce yeasts that mount to the surface during the barrel aging. These yeasts that naturally occur in the grapes and the barrel form a cap known as flor.

This yeast cap, which feeds on the wine while protecting it from oxidation, confers on fino and manzanilla sherries their distinctive flavors of almonds and walnuts and conserves their primary aromas, among which are green apple and olives, as well as a touch of saltiness in the case of manzanilla.

These wines, generally dry, even if some are quite sweet, such as Pedro Ximénez and sometimes oloroso, are classified in different categories. They go from the youngest, pale and fresh finos and manzanillas, raised in the absence of oxygen, to the older ones: colorful, complex amontillados, palo cortados, and olorosos, whose maturation takes place in the almost constant presence of oxygen.

There are also several minor style variants, such as cream sherry, a very sweet oloroso; manzanilla pasada, an amontillado from Sanlúcar de Barrameda; and Pedro Ximénez, a grape variety that makes a very sweet, concentrated sherry.

It is worth noting that in Montilla-Moriles, all wines are based on Pedro Ximénez grapes, even dry wines such as finos and amontillados. The wines from Jerez are based on the Palomino grape, except of course those sherries whose labels are marked Pedro Ximénez, sometimes abbreviated as PX.

### AN IMMENSE AROMATIC POTENTIAL

Presently, some 307 volatile compounds have been identified in the different types of sherries and Montilla-Moriles wines. Our olfactory capabilities enable us to identify only a certain proportion of volatile compounds, generally those with the greatest olfactory power.

Several of these compounds are also present in other categories of wines (dry and sweet white wines, red wines, rosé wines, and sparkling wines) as well as in many foods. These include acetals, acids, alcohols, amides, bases, carbonyls, sulfur-containing compounds, coumarins, dioxolanes and dioxanes (which are acetals), esters, furans, lactones, and phenols.

## THE FORMATION OF AROMATIC COMPOUNDS

The formation of sherry's aromatic compounds occurs during several stages of its development. The production of these compounds is affected by the choice of grape varieties planted, growing conditions (terroir, climate, etc.), fermentation, and slow maturation, which occurs underneath a yeast cap (especially for fino sherry) or in a more oxidative milieu (for oloroso and sometimes amontillado). Aromatic compounds also come from adding alcohol and from the impact of oak's volatile compounds, which dissolve into the sherry during barrel maturation.

Sugar decomposition (glycolysis) over time, occurring in creamy and syrupy wines as well as fortified wines such as sherry, also generates new aromatic compounds.

## FINO AND MANZANILLA: ANDALUSIAN FRESHNESS

Fino sherries are developed from wines based on Palomino grapes. These are the palest and lightest sherries, originally containing between 11% and 12% alcohol. The press juice is rarely used for fino. Once the fermentation process is finished, the sherry is fortified to 15.5% with neutral eau de vie, which does not add any new aromas to the wine but does have an impact on the development of volatile compounds.

## ALCOHOL'S IMPACT ON FRAGRANCES

As is the case for port, adding alcohol to sherry, raising its level to between 15% and 20%, and stopping fermentation has an impact on its aromatic compounds, in particular inhibiting some fragrances. On the other hand, increasing sherry's alcohol level after fermentation and before its maturation in the *solera* increases the extraction of aromatic compounds such as phenols from the oak. This phenomenon is particularly marked for oloroso sherries.

Fino sherry is conserved in barrels located in the coldest part of the cellar to help the development of the yeast cap, or flor, which prefers coolness to heat, explaining its particularly high level of activity from February to June. In addition to augmenting the growth rate of yeast, the temperature affects its metabolism, the quantity and nature of the metabolites, and the chemical transformations carried out by the yeast enzymes, which affect fino sherry's final aroma. A few degrees' difference is enough to produce organoleptic variations in foods and wines.

The barrels, which are filled to only 80% of capacity in order to promote the growth of flora, are organized in *solera,* which, as previously mentioned, is an aging system involving a combination of older and newer wines arranged in barrels.

The microorganisms in the yeast cap consume oxygen and thus protect the wine against oxidation and the browning of its phenolic compounds. This is the reason for fino's coolness and very pale yellow color.

In contrast, oloroso is not protected by a yeast cap. The oxidation of its phenolic compounds gives it a dark color and a more evolved bouquet with grilled aromas.

During fino's maturation under the yeast cap, the level of many fatty acids and aromatic acids increases, as does the level of terpenes and carbonyls. More than thirty-six new aromatic compounds come from the vinification under the yeast cap, including fourteen acetals, two acids, three alcohols, four carbonyls (aldehydes and ketones)—ketones, which are typically very volatile, are, at least in theory, the first to reach the olfactory centers and should be sensed at the initial smelling—four esters, one lactam, four lactones (among the most distinctive fino compounds), and four nitrogen compounds.

Fino's principal esters, ethyl acetate and ethyl lactate, increase right from the start of the maturation process occurring under the yeast cap, only to see their concentrations diminish at the end of the process.

### THE METABOLIC ACTION OF THE YEAST CAP

The yeast cap so characteristic of Andalusia provokes numerous biochemical changes and determines the wine's final character. There is a decrease in alcohol (about 1.6%) and a more pronounced reduction in glycerol and volatile acidity—the

ESE

<br>

yeast uses all three of these things as carbon sources for the development of the cap, which explains their decreased quantities.

There is also an important increase in acetaldehydes (green apple fragrance), in particular at the beginning of the maturation process under the yeast cap. The acetaldehydes, along with the lactones, largely determine fino's aromatic profile, giving it green apple and walnut tonalities.

Acetaldehydes are also the precursors of many other volatile compounds, such as dimethylketol and 1,1-diethoxyethane—the latter is also present in apricots and some cherry varieties. They react with selected phenolic compounds and alcohols, generating new aromatic compounds, some of which play a large part in fino's specific characteristics.

Another factor in fino's aromatic singularity is the autolysis of dead yeast cells; in other words, the spontaneous rupturing of yeast cell membranes that then release their contents, nourishing the lees of these *sous-voile* wines.

### MANZANILLA

Manzanilla is a regional variant of fino, developed from grapes cultivated on the Mediterranean hills of the village of Sanlúcar de Barrameda. This generally leads to a lighter fino than the one carrying the appellation sherry, with sharper bitter and iodine tonalities. The cooler climate and humidity explain the stylistic difference between manzanilla and fino—even though the ocean's proximity is often invoked to explain manzanilla's iodine character. Manzanilla pasada is an amontillado sherry coming from the same village.

### FINO'S GOURMET MOLECULES

The dominant volatile compounds of fino and manzanilla have the potential to lead us to inspiring paths for harmonious food and wine pairing. These include acetaldehydes (also present in walnuts, green apples, and Spanish ham), acetoin (fatty, creamy, and buttery flavors, such as in butter and yogurt), lactones (apricot, peach, coconut), diacetyls (butter and cheese), solerone (dried figs), and terpenes (citrus fruits and flowers).

Acetoin (dimethylketol), which imparts a fatty, creamy, and buttery flavor, is one of the most important volatile compounds of fino and manzanilla, but it also helps provide an aromatic

identity to other foods. It is found in fresh or cooked apples, fresh or cooked leeks, asparagus, broccoli, Brussels sprouts, roasted coffee beans, strawberries, quinces, cantaloupe, corn syrup, fermented tea, and of course butter, cheese, milk, and yogurt. These last three are particularly rich sources.

All of these ingredients, because of their molecular compatibility, can be combined to create harmonious recipes, especially when paired with fino and manzanilla sherry.

Two other volatile compounds also heavily contribute to the aromatic uniqueness of these wines. First, there are the lactones, or more precisely the sherry lactones, also occurring in apricot, coconut, peach, pork (including Spanish ham), and vanilla. Interestingly, one of these sherry lactones, solerone, a close relative of sotolon (see the chapter of that name), also characterizes the aromatic profile of dried figs, as well as that of dates and smoky, black Lapsang Souchong tea.

The pairing of dried figs with fino illustrates the foundations of this research on the aromatic molecules in foods and wines in the search for wine and food pairing possibilities.

### DRIED FIGS AND FINO SHERRY: A REVELATION!

For a long time, I was convinced that amontillado and oloroso sherries were the best companions for figs. They are already richly endowed with the aromatic tonalities often accompanying dried fruit. Furthermore, their amber, brownish color is reminiscent of dried figs.

But to my great surprise, I discovered that fino sherry, with its very fresh aromatic notes and pale yellow color, is *the* made-to-order companion for dried figs. It easily dethrones amontillado and oloroso. This, despite fino's more imposing dryness in the mouth, given that it has a lower level of glycerol than amontillado and especially oloroso.

The harmony here is more than ever molecular, because the sherry and the figs complement each other in the mouth (the wine lightens the fig's sugar), and especially because of the meeting of the solerone molecules found in both fino and dried figs.

In fact, dried figs and fino sherry share several volatile compounds (aldehydes, phenols, and lactones), among which solerone is one of the most important aromatic compounds. This volatile compound, typical of wines raised under a yeast

cap, is essential. Solerone (which is, as I mentioned before, a close relative of sotolon) is a member of the lactone family, found more often in fino sherry than in amontillado and oloroso, which do not mature under a yeast cap, or do so only slightly. Therefore, solerone, the aromatic key of dried figs, finds a friendly "lock" in fino sherry!

Very floral terpenes, smelling of lavender and lilies of the valley, are also found in a wide variety of foods that can become complementary ingredients: sweet European basil, bergamot, rosewood, Ceylon cinnamon, coriander, fresh figs, ginger, mint, nutmeg, olives, grapes, rice, rosemary, saffron, and citrus zests.

Now that you're familiar with this very large family of foods possessing a substantial power of attraction, you can use your newfound knowledge to experiment in the kitchen, as well as at the table, when you add fino or manzanilla sherry into the mix. Choose some ingredients from the lists in the accompanying figures and create new recipe ideas. Here are a few to start you off:

+ Spanish ham (or prosciutto) plate with dried figs (instead of the traditional cantaloupe)
+ Skewers of dried figs rolled in Spanish ham (or prosciutto)
+ Half-ripened goat cheese macerated with rosemary
+ Creamed aspic of leeks and apples (fresh and cooked) spiced with fresh coriander
+ Grilled scallops and couscous with Brazil nuts and sprigs of grilled coconut, ginger, and citrus fruits in a yogurt sauce

**SOMMELIER-COOK'S HINT**

**Rosemary cheese and fino sherry** Fino and manzanilla sherries are quite rich in floral, terpenic notes (arising from different aromatic molecules like linalool, nerolidol, and farnesol), which makes them good partners for rosemary, especially because their aromatic power and presence in the mouth stand up to rosemary's own expressivity. One excellent combination is fino or manzanilla with a dry goat cheese salad in which the cheese has been marinated in rosemary-flavored olive oil.

## MEALS WITH FINO AND MANZANILLA

Even though they are almost always served as aperitifs, fino and manzanilla sherries deserve a special place at your table. I would even go so far as to say that they can be considered among the most versatile wines when it comes to food pairings.

Are you skeptical? You may think that it's impossible that a single wine could successfully support the salty, iodized flavors of oysters, crustaceans, and caviar, as well as the complex tastes of sushi and its condiments (soy sauce, marinated ginger, daikon, and wakame), and yet also easily accompany green asparagus, olives, artichokes, smoked fish, and ripened goat cheeses.

Although it is indeed a rare pearl, the king of food pairings really exists, and it's fino sherry. With its 307 identified volatile compounds, like manzanilla, its royal companion, fino offers a very wide palette of aromas that can complement a vast array of ingredients.

## A NEW CREATION AT EL CELLER DE CAN ROCA, INSPIRED BY CHARTIER

Over the course of various trips to Spain, I have worked both with sommeliers practicing molecular sommellerie and with chefs aware of molecular wine and food pairings, to help develop recipes exploiting my research on the aromatic structure of food. I'd like to share one of these lovely compositions, created by a great Catalan chef.

The distinguished sommelier Josep Roca, co-owner of the restaurant El Celler de Can Roca—which has three stars from Michelin, was designated the fifth best restaurant in the world in *Restaurant Magazine's* 2009 list of the Top 50, and is a leading European light in food and wine pairing—participated in

my workshop at elBulli in September 2008. He was inspired by my work, and with his brothers, chefs Joan and Jordi Roca, he created a dish in homage.

Thus, added to the specialities of this famous Girona restaurant was *"Homenaje a François Chartier/Cigala en tempura de almendras tiernas, compota de manzana, curry u hongos"* (Homage to François Chartier/Langoustines in a tempura of grilled almonds, curried applesauce, and mushrooms).

This dish was a result of my research on the aromatic compounds of fino sherry, a centerpiece of the Spanish wine industry, and the foods of the same aromatic family presented in this chapter. Acetaldehydes, solerone, and sotolon, as well as acetoin, terpenes, and lactones, are at the heart of the aromatic molecules of this harmonious creation by the Roca brothers.

## OLOROSO: THE CARESSING SWEETNESS OF SHERRY

Oloroso sherry is developed from wines that are darker and richer in phenolic compounds than those of fino sherry; in fact, oloroso has more than twice the phenols of fino. It also has almost twice the volatile acidity of fino sherry.

In this case, we are not talking about maturation under a yeast cap. The base wine is fortified to reach 18.5% to 20% alcohol; this alcohol effectively inhibits the yeast that would normally create a cap.

Oloroso barrels are filled to 95% of their capacity, in contrast to 80% for fino, and they are placed in the warmest part of the cellar—sometimes even outside the cellar, where they bask in the Andalusian sunshine and heat—to promote the oxidation of the phenolic compounds. The resulting product is dark brown, with aromas that are more developed and more roasted than those of fino. Oloroso sherry also has a denser body than fino; it's full and enveloping, in part because of its rich glycerol content. In contrast, fino sherry has lost its glycerol because of the yeast cap. This explains in part its more compact and slightly bitter, very dry profile.

The higher alcohol level of oloroso (18.5% to 20%) helps extract phenolic compounds from the wood (phenols are alcohol soluble) to generate a more aromatically complex sherry during its development in an oxidative milieu. Phenols (which include benzoic acid, cinnamic acid, phenolic aldehydes, and coumarin) largely contribute to the aromas and flavors of wine.

# 1. VOLATILE AND AROMATIC COMPOUNDS
### FINO AND MANZANILLA SHERRY

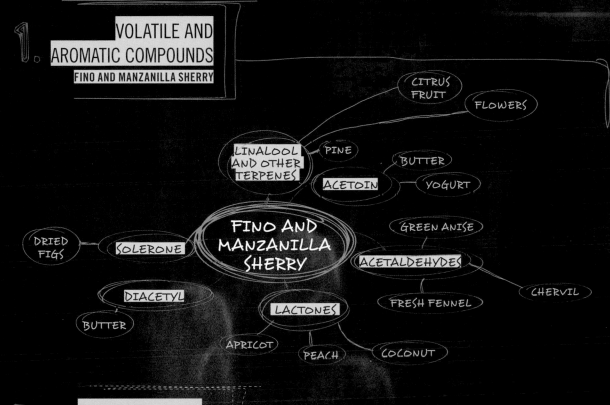

CITRUS FRUIT

FLOWERS

LINALOOL AND OTHER TERPENES

PINE

BUTTER

ACETOIN

YOGURT

FINO AND MANZANILLA SHERRY

GREEN ANISE

DRIED FIGS

SOLERONE

ACETALDEHYDES

CHERVIL

FRESH FENNEL

DIACETYL

LACTONES

BUTTER

APRICOT

PEACH

COCONUT

# 2. COMPLEMENTARY FOODS
### FINO AND MANZANILLA SHERRY

APPLES
APRICOTS
ASPARAGUS
BERGAMOT
BLACK SMOKED TEA

BROCCOLI
BRUSSELS SPROUTS
BUTTER
CANTALOUPE
CEYLON CINNAMON
CHEESE

CITRUS FRUIT
CITRUS ZESTS
COCONUT
CORIANDER
CORN SYRUP
DATES
EUROPEAN SWEET BASIL
FRESH AND DRIED FIGS
GINGER
GRAPES
GREEN APPLES
LAVENDER
LEEKS

MILK
MINT
MUSHROOMS
NUTMEG
NUTS
OLIVES
PEACHES
PORK
QUINCES
RICE
ROASTED COFFEE BEANS
ROSEMARY
ROSE WATER

ROSEWOOD
SAFFRON
SPANISH HAM
VANILLA
YOGURT

Oloroso, whether dry or sweet, always leaves the impression of being sweet and syrupy because of its richness in alcohol (ethanol) and in glycerol, both of which create the sensation of roundness, texture, and sweetness (without sugar).

Oloroso spends some eight to twelve years maturing in barrels. After twelve years, it has lost 30% to 40% of its initial content. This concentration of ingredients greatly helps in the development of new volatile, aromatic compounds and creates a denser, thicker texture than that of fino and amontillado.

## MEDIUM DRY, MEDIUM, CREAM, OR PALE CREAM SHERRY?

Generally, these different types of sherry are developed from an oloroso base to which a touch of fino is added, rendering its color paler and adding aroma, plus a touch of amontillado to make its bouquet more complex.

## OLOROSO'S AROMATIC COMPOUNDS

Oloroso, much like amontillado but to a lesser extent, is richly endowed with phenolic compounds. These originate in the base wine selected for this type of sherry, as well as in compounds extracted from the oak in the barrels. Oloroso and amontillado sherries are rich in phenols compared to fino; they have a higher level of alcohol and alcohol is a solvent.

Four phenolic compounds are dominant in oloroso: benzoic acid (almond aroma), cinnamic acid (cinnamon aroma), phenolic aldehyde (walnut aroma), and coumarin (vanillin, tonka bean, and cut hay aromas).

Oloroso's aromatic signature is, as I described previously, marked by several aromatic compounds, such as solerone, but to a lesser degree than for fino sherry.

We also find in oloroso the very aromatic and complex sotolon (aromas of curry, walnuts, roasted fenugreek seeds, maple syrup, and soy sauce), sherry lactones (aromas of dates, dried figs, coconut, black smoked tea, and vanilla), and piperonal (sweet and floral aromas).

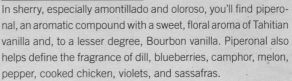

## PIPERONAL (HELIOTROPIN)

In sherry, especially amontillado and oloroso, you'll find piperonal, an aromatic compound with a sweet, floral aroma of Tahitian vanilla and, to a lesser degree, Bourbon vanilla. Piperonal also helps define the fragrance of dill, blueberries, camphor, melon, pepper, cooked chicken, violets, and sassafras.

## OLOROSO AT THE TABLE

Given its structure, which is characterized by ampleness, warmth, density, fat, sweetness (with or without sugar), and powerful aromas, oloroso—affected by the liberation of high molecular density aromas—necessitates above all textured and fatty foods, which can deal with bitterness. Ideally it should be served at a moderate temperature so that the heat does not exacerbate the alcohol.

It is also important to choose or create dishes dominated by ingredients whose aromatic signatures carry these same volatile compounds. For example, oloroso contains coumarin, which also appears in broad beans, lavender, cloves, cinnamon (mostly Chinese cinnamon or cassia cinnamon), licorice, angelica, tobacco, and artificial vanilla extract, as well as bison grass, which is used in Polish Zubrowka vodka.

## SOMMELIER-COOK'S HINT

Have fun in the kitchen with ingredients complementary to oloroso sherry by making a dessert with cocoa, roasted fenugreek seeds, dates, and Tokaji Aszú (a sweet Hungarian wine). If you want to go salty, prepare a "smoked" fish using Lapsang Souchong tea. Steep this smoked black tea in a heated cream sauce just prior to serving, and coat your unsmoked fish with it, creating the illusion of smoked fish! Finally, if you have an elBulli-style food siphon, don't hesitate to serve smoky foam with your fish nuggets. You'll get the same results and obtain real harmony with an oloroso sherry that's dry rather than sweet.

## AMONTILLADO

At the onset, amontillado is a wine that is vinified under a yeast cap, as is fino, and is fortified at a mere 15.5% alcohol. But during its evolution in the same *solera* used for fino, where the barrels are filled to only 80% of capacity, this cap disappears.

## 3. VOLATILE AND AROMATIC COMPOUNDS
### OLOROSO SHERRY

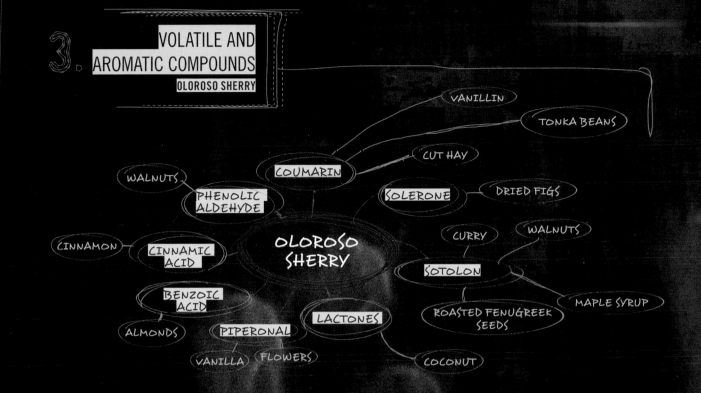

OLOROSO SHERRY

- VANILLIN
- TONKA BEANS
- COUMARIN
  - CUT HAY
- WALNUTS
- PHENOLIC ALDEHYDE
- SOLERONE
  - DRIED FIGS
- CINNAMON
- CINNAMIC ACID
- CURRY
- WALNUTS
- SOTOLON
- BENZOIC ACID
- MAPLE SYRUP
- ALMONDS
- PIPERONAL
- LACTONES
- ROASTED FENUGREEK SEEDS
- VANILLA
- FLOWERS
- COCONUT

## 4. COMPLEMENTARY FOODS
### OLOROSO SHERRY

ALMONDS
ANGELICA ROOT
BLACK SMOKED TEA
CELERY SALT
CHINESE CINNAMON
CLOVES
COCOA
COFFEE
CURRY

DATES
DRIED FIGS
FLAX OIL AND FLAXSEEDS
HAVANA TOBACCO LEAVES
LICORICE
MAPLE SYRUP
POLISH ZUBROWKA VODKA
ROASTED FENUGREEK SEEDS
SAKE
SMOKE

SMOKED FISH
SMOKED MEAT
SOY SAUCE
TOASTED COCONUT
TOKAJI ASZÚ
TONKA BEANS
VANILLA
VARIOUS PEPPERS
WALNUTS

The wine is once again fortified to reach between 17% and 18% alcohol, and then it is transferred into another *solera* whose barrels are filled to 95% in an oxidative milieu, thus without a yeast cap.

Over time, the wine's pale initial color becomes amber. The fino aromas of apple and fresh almonds are affected by the controlled oxidation, transforming into the notes of grilled walnut that are characteristic of amontillado. The wine becomes more full-bodied and more generous than fino as its glycerol level increases.

Amontillado is situated halfway between fino and oloroso, both in terms of aroma and in terms of pairing options.

## WHISKY LACTONES AND DATES

Scotch aged in sherry barrels (such as Macallan 10 Years Old Fine Oak and Macallan 10 Years Old Cask Strength) also contains lactones, including the famous whisky lactones and sherry lactones, like the solerone of sherry and dried figs.

As is true for dried figs, dates are richly endowed with phenolic compounds. This enables the development of the same kinds of aromas, thanks to the browning (oxygenation) of the phenolic compounds and the Maillard reaction taking place among the sugars and amino acids (dried dates are very rich in sugar, which represents approximately 60% to 80% of their weight).

We can therefore envisage pairings with wines characterized by the same aromatic molecules as those found in dried figs.

SOMMELIER-COOK'S HINT

**Lactone recipe** Be daring and create a recipe based on dried figs and Scotch aged in sherry barrels, with vanilla and toasted coconut (two aromas in the lactone family). This can pair perfectly with a fino or manzanilla sherry, or an amontillado initially raised under a yeast cap that finished maturing in an oxidative milieu.

You may add dates to this recipe, as they are in the same pairing family as dried figs; both are rich in phenolic compounds. Because sotolon, another member of the lactone family, is a close relative of solerone, you may pair it with a botrytised wine, such as a Sauternes or a Tokaji Aszú, that is over ten years old. Other possibilities include using maple syrup and roasted fenugreek seeds in desserts.

SMOKED FISH (AND SIDES OF SALMON
COOKED ON ONE SIDE)
APPLES
POWDERED MALT

SMOKY

GUAIACOL

BARREL-RAISED
WINES

# OAK AND BARRELS

## OAK: THE MAESTRO OF AROMAS AND FLAVOR ENHANCEMENT

"Musical scores tell everything except the essential."

GUSTAV MAHLER

It's a well-known fact that oak barrels are a central element in the development of great wines and brandies, and have been for a very long time. Even though they were already being enjoyed at the time of the Gauls, it was towards the end of the nineteenth century that the organoleptic qualities (that is to say, qualities appealing to our sensory organs) of barrels were really recognized.

Contemporary written tasting notes of wine brokers of that period showed that certain wine aficionados were charmed by the top-of-the-line Burgundies that were raised in new oak barrels...

What followed is well known. At one point or another, most of us have taken advantage of the benefits of barrels for wine and unfortunately have also occasionally suffered from their misuse in wine production.

I will not present here a comprehensive study of oak; this subject has already been covered extensively in the viticultural literature. Instead, my goal is to try to ascertain its aromatic impact on wine and brandy, with the goal of establishing an aromatic map of some wines and brandies that were nurtured in oak. That way, we can more easily make connections between these oaked beverages and their complementary ingredients.

### AROMAS BORN IN FIRE

Before oak barrels are filled with wine, their interiors are burned (toasted) with a wood fire. The flames, of varying intensities, actually come into contact with the wood. The process serves to soften the naturally bitter and rough structure of the oak tannins to obtain a suppler product and to transform the oak's naturally occurring volatile compounds.

The combustion of a food or organic matter such as wood sets off multiple chemical reactions (oxidative reductions), essentially between the oxygen in the air and the molecules, generating aromatic molecules possessing distinct physico-chemical and organoleptic properties.

One must also consider the alcohol content of barrel-raised wine or brandy because this will affect its solubility (physico-chemical property) and thus the amount of specific molecules in the final product. For example, phenols such as guaiacol (which smells like smoke) and eugenol (which smells like cloves) are slightly water soluble, but very soluble in ethyl alcohol.

Therefore, the guaiacol within a barrel will be extracted more efficiently if it is used for aging a Scotch rather than wine. The alcohol also plays the role of a transporter, moving the less volatile aromatic molecules such as maltol (which smells like burnt sugar) towards the olfactory centers.

The scents of untoasted new oak are substantially affected by the volatile compounds expressed in notes of vanilla, nutmeg, coconut, spices, and leather, as well as by vegetal and earthy aromatic tones.

Once fire-roasted, the oak barrels develop new aromatic compounds. The decomposition of the oak's lignin gives rise to

BALSAMIC VINEGAR
BROWN SUGAR
BUTTER
CARAMEL
CARDAMOM
CELERY
CHICORY
CINNAMON
CLOVES
COCOA
COFFEE
COTTON CANDY
CURRY
EUCALYPTUS
FOODS COOKED ON A WOOD-
FIRED OR CHARCOAL GRILL
FOODS RICH IN UMAMI
(SCALLOPS, SEAWEED, CHEESE, BRAISED MEAT,
SHIITAKE MUSHROOMS)
GRILLED ALMONDS
GRILLED BEEF
GRILLED OR ROASTED ASPARAGUS
HAVANA TOBACCO LEAVES
LEMONGRASS
MUSHROOMS
PINEAPPLE
QUINCES
SANDALWOOD
SPICES
TONKA BEANS

METHYLOCTALACTONES

GRILLED ALMONDS

FURANIC
ALDEHYDES

MAPLE
SYRUP

VANILLA

GRILLED,
ROASTED, AND
TOASTED

VANILLIN

CYCLOTENES

VANILLA

WINES RAISED

several very volatile phenolic aldehydes, including vanillin, the major source of the more or less strong vanilla odor and of the disappearance of vegetal notes.

This process generates several other phenols, with smoky notes, such as guaiacol, and woody and spicy notes in the case of eugenol, the main aromatic compound of cloves. There are also strong methyloctalactones, better known as whisky lactones, with their toasted aromas found in coconut and elsewhere.

Other compounds that appear include furanic aldehydes (roasted almonds), furfurals (aromas of sawdust, maple syrup, toast, roasted almonds, coffee, and caramel), and 5-methyl-furfural (roasted almonds).

You'll also find cyclotene (maple syrup, grilled and roasted licorice), maltol, and isomaltol (with notes of burnt sugar and caramel). These last three are characterized by much more powerful caramelized and toasted tones than are the furfurals.

The aromatic tonality of toast comes equally from its cyclotene and maltol, two volatile molecules generated when oak is subjected to fire. Cyclotene and maltol are also partially responsible for the fragrance of toast and cookies that can be detected in beer, cheese, and pastries.

The aroma of chocolate, also generated when oak barrels are heated, comes from the presence of aldehyde derivatives, such as acetylpyrrole.

Several other volatile wood phenols are also released into the wine during its maturation in toasted barrels, such as 4-methylguaiacol, 4-propylguaiacol, and 4-ethyl-2,6-di-methoxyphenol, as well as other compounds like phenyl-ketones and aldehydes.

## THE AROMATIC COMPOUNDS OF TOASTED WOOD AND SMOKE

When wood burns, whether in a wood stove, a charcoal grill, or during the toasting of oak barrels, a multitude of new aromatic compounds permeate the wood and the ensuing smoke.

One can detect families of volatile compounds such as furans (with a sugary fragrance of bread and flowers), lactones (coconut), and phenols (astringent and smoky).

Also found are powerful aromatic molecules such as ethanal or acetaldehyde (green apples), 2,3-butanedione or diacetyl (butter), 2-methoxyphenol or guaiacol (smoke), 2-methoxy-4-formylphenol or vanillin (vanilla), 4-allyl-2-meth-oxyphenol or eugenol (nutmeg), and 2,6-dimethyloxyphenol or syringol.

Syringol is a phenol that comes from the burning of wood (lignins or vegetal walls themselves composed of phenols), resulting in spicy, smoky, vanilla, and medicinal aromas. It also naturally occurs in onions, garlic, leeks, and shallots, which upon cooking (when their sugars caramelize) become very interesting bridging ingredients to some wines matured in oak.

There are close relationships between the aromas of wines that have spent time in barrels and the aromas of charcoal-cooked food.

As I mentioned previously, vanillin is produced when oak barrels are toasted. This partially explains the vanilla fragrances found in wines, Scotches, and cognacs and in all foods cooked or smoked on wood fires, including bread, fish, some vegetables, and meat. These foods also develop caramelized/spicy/roasted notes from their contact with the smoke released during cooking. This explains the almost perfect harmony between oaked wine and grilled food.

## THE MAILLARD REACTION

The Maillard reaction occurs in sugars and amino acids when the temperature rises sharply, such as during the cooking of meat or the toasting of the insides of oak barrels. These processes generate colored, aromatic substances (melanoids) that can be found in grilled meat and the crust of toasted bread, as well as in wine. In oak, the sugars and amino acids found in wood, once heated, are transformed into empyreumatic notes including caramel, cocoa, toast, vanillin, and ethyl vanillate (a vanillin derivative).

## AROMATIC DIFFERENCES DEPENDING ON THE OAK'S ORIGINS

The varieties of oak found in central France and Russia are the richest in eugenol (the compound found in cloves). This opens the door to a very large palette of wines that provide, to varying degrees, the fragrance of cloves, and can thus be paired with this spice, as well as with other foods rich in eugenol (see the "Cloves" chapter).

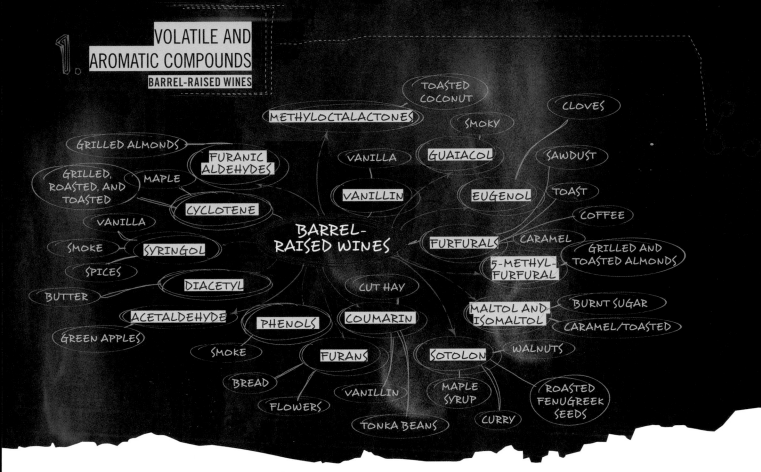

American oak, which also contains eugenol, but in a smaller amount than the two varieties mentioned above, is particularly affected by methyloctalactones (whisky lactones). This is also the case for French oak from the Vosges forests, whose dominant fragrance is coconut.

For their part, Russian oaks and oaks from the forests of the Centre region of France, which are dominated by the Sessile oak *(Quercus petrae)*, have two or three times as much aromatic impact as do those from the forests of Burgundy and Limousin. These forests are dominated by the Pedunculate oak *(Quercus robur)*, which has a greater physical presence in releasing more tannins and other polyphenols.

This explains why wine producers use the latter oak more frequently for aging brandy, while oaks from the Centre and Vosges regions of France, with their softer, less abundant tannins, are better for aging wines.

It is interesting to note that the lignins present in the oak used to make barrels are composed of the same molecules as the lignins in grape stalks. This helps explain the natural link between the oak barrel and wine, which is marked by the presence of compounds originating in the grape stalks.

## AMERICAN OAK

American, Australian, and Spanish wines are often partially or completely aged in barrels made from American oak (*Quercus alba* in particular). These barrels, whose insides are often much more highly toasted than their European counterparts, infuse the wine with many aromatic molecules, such as the coconut-scented methyloctalactones (whisky lactones).

The family of lactones produces complex milky notes that may also show strong touches of vanilla, balsam, wood, sugar (pastries), or earthiness, with notes of apricot, peaches, leather, spices, green walnuts, or fresh herbs.

Given that coconut is more heavily present in wines raised in American oak barrels, one must consider the perfect union between dishes dominated by coconut (whether based on fresh coconut, toasted coconut, or even coconut milk) and oaked Chardonnay, which brings out in the wine milky tastes reminiscent of coconut. On the other hand, it is important to note that when white wines are barrel-fermented and mature on their lees (sediments), this process reduces vanillin's aromatic impact because it is transformed into scentless vanillin alcohol.

Chardonnay raised in oak barrels and coconut-based recipes really go well together! These foods also pair well with any wines raised in American oak barrels because they contain touches of caramel, cocoa, and toast, caused in part by the Maillard reaction. Caramel and chocolate are usually more noticeable in wines raised in American oak. As for the aroma of toast, coming from maltol and cyclotene, this aroma is readily perceptible even in very diluted quantities.

## OAK CHIPS

Aldehydic phenols, such as vanillin (vanilla), are found in large quantities in red and white wines raised in the presence of toasted oak chips.

On the other hand, volatile phenols, such as maltol (burnt sugar/cotton candy) and eugenol (cloves), are usually found in smaller quantities in wines that have come into contact with oak chips. It is thus necessary to seek out pairings based on smoked foods or chocolate, vanilla, or tonka beans to accompany wines that were raised with oak chips rather than in barrels.

### A FLAVOR ENHANCER

Maltol, with its aroma of burnt sugar, is a naturally occurring organic compound often found in oak, larch, pine, and licorice, as well as in roasted malt produced by the Maillard reaction. It also occurs generously in wines raised in oak barrels, as well as in beer and Scotch, both of which are based on roasted malt, which is also quite rich in maltol. Furthermore, Scotch spends many years in oak barrels. Maltol is also a flavor enhancer used in the food industry. This is the chemical that gives cotton candy its unique fruity/caramel taste!

### THE AROMATIC PROFILE OF BARREL-RAISED WINE

Eugenol (cloves) is one of the principal aromatic signatures of toasted oak barrels, and one of the most easily recognized—at least to my nose! It particularly dominates wine raised in oak barrels from the Centre region of France, as well as in oak barrels from Russia.

Another important aromatic signature of oak barrels is vanillin, probably the most recognizable compound to most wine drinkers. Vanillin occurs in all the varieties of oak used to make barrels, once the oak has been toasted. However, after strong toasting, it is American oak that usually has the highest level of vanillin.

Vanillin and eugenol are sometimes present in trace amounts in wine that has not yet been barreled. However, their concentration increases considerably as the wine ages in the barrel.

Among the most important compounds that wine develops during barrel maturation are methyloctalactones. Also known as whisky lactones, these are the most characteristic volatile oak compounds, whose aroma includes that of coconut. Actually, it is the *cis* isomer of methyloctalactones, four times more aromatic than the *trans* isomer of the same molecule, that is found in all oak and that is *the* trademark aroma of barrel-raised wine.

As I mentioned previously, these methyloctalactones present themselves in a complex milky fragrance of coconut, which can also include vanilla, balsam, wood, sugar, or earth notes, plus touches of apricot, peach, leather, spice, and unripe walnut or fresh herbs.

It is American oak—which also contains eugenol (cloves), but in a lesser quantity—that has the largest amount of methyloctalactones (coconut). This is also true for French oak from the forests of the Vosges. Note that it's necessary to toast American oak for a longer time in order to curb its aromatic impact. As a result, wine raised in American oak will contain more vanillin.

Lactones, smelling of coconut, apricots, or peaches, and which may have tones of vanilla, balsam, wood, or sugar, heavily transform wine's complex aromas during their barrel maturing.

Coumarin, a lactone, is another aromatic compound that comes from oak. It has complex fragrances of vanilla, tonka beans, and cut hay, and an acidic, bitter taste. Given that it is one of the aromatic compounds of licorice, it helps provide a licorice-like taste to the tannins associated with red wines matured in new oak barrels. This is why I sometimes use the term "licorice-like tannins" in my wine descriptions.

Coumarin is also one of the major volatile compounds in lavender, cloves, cinnamon, angelica, tonka beans, tobacco, artificial vanilla extract, bison grass (a plant used in making Polish Zubrowka vodka), and some very peaty Scotches, such as the ten-year-old Talisker from the Isle of Skye.

Furfurals—volatile compounds contained, for example, in sawdust and maple syrup—reach high concentration levels in wine raised in oak barrels. Their penetrating fragrances play a role in creating complex tonalities that simultaneously have sugary, woody, caramel, and hazelnut characteristics with nuances of toast, burnt food, coffee, and astringency.

To be more precise, furfural is a direct precursor of 5-furfuryl-mercaptan, which appears in strong concentrations during barrel fermentation, in particular during malo-lactic fermentation (the transformation of the harsher malic acid into the softer lactic acid) and also during the maturation of wine on its lees. This quite aromatic component, the principal element in espresso coffee's fragrance, is absent from oak, but is created in barrel-raised wines when the oxidation level is low. The same process occurs for methyl-5-furfural, creating 5-methyl-furfuryl mercaptan, which smells of toast and coffee. These compounds are found, for example, in the fragrance of Champagne conserved on its lees and in the fragrance of Sauternes. In fact, the oak serves as an extender and amplifier for these aromas in red wines and certain white wines.

We also find, but to a lesser extent, beta-caryophyllene, a molecule with a wood fragrance and furanic compounds, which are molecules formed when sugar is caramelized. It is associated with the aromas of almonds, roasted almonds, and toast.

## CLOVES AND WINE

Eugenol is the dominant aromatic compound in cloves, but it also appears in Thai basil, wild basil, malted barley, fresh mangos, apricots, pineapple, strawberries, cinnamon, rosemary, potatoes, cooked asparagus, mozzarella cheese, and grilled beef.

Because cloves contain traces of aromatic compounds such as furfural (a molecule also found in sawdust, among other places), beta-caryophyllene (a compound with a woody fragrance that is also found in black pepper), and vanillin, there is a strong aromatic bridge between cloves and wines that matured in oak barrels, particularly barrels made of oak from the forests of Centre, France, and of Russia, as these are the varieties that are richest in eugenol.

Given their very powerful, sensual fragrance, cloves work very well with rich, penetrating red wines, in particular some Spanish vintages that are rich in eugenol.

Their best companions are Spanish reds based on Garnacha (Grenache) from the Priorat, Rioja Baja, Campo de Borja, and Cariñena regions. Cloves also pair very well with Tempranillo-based wines from the La Rioja, Toro, and Ribera del Duero regions, and with wines made from the Mencia grape grown in Bierzo. Portuguese wines based on the Touriga Nacional, a grape variety from the Douro region, are also a good choice, as are many Pinot Noirs, whether from Burgundy or New Zealand.

## MAPLE SYRUP AND OAK BARRELS: THE SAME AROMATIC PROFILE

Oak and maple are two essences that, when transformed by heat—oak barrels are toasted on the inside before their use, and maple water is heated at an elevated temperature to be transformed into syrup—see their volatile compounds, stemming from the wood lignins in oak and maple, get magnified into a collection of new molecules that are even more complex and more aromatic.

Compounds common to oak-matured wines and maple syrup (see the "Maple Syrup" chapter) include lactones (milky [coconut], fruity [apricot/peach], tasting of nuts [almond/hazelnut] or of caramel), furanones (in particular maple furanones, smelling of caramelized maple), maltol (burnt sugar), cyclotene (whose strong fragrance is situated between maple and licorice), and furfural (with sugary, woody, toasty, hazelnut, and caramel fragrances).

In addition, let's not forget the famous vanillin; the powerful eugenol; 4,5-dimethyl-3-hydroxy-2(5H)-furanone, a.k.a. sotolon (a complex fragrance with touches of walnuts, curry, toasted fenugreek seeds, caramel, and maple syrup); syringaldehyde (whose fragrances are discretely smoky and chocolatey, reminiscent of vanillin and tonka beans); the subtle beta-caryophyllen (woody fragrances); as well as several other molecules present in oak and maple wood.

## OAK AND MAPLE: A SURPRISING AROMATIC CONNECTION!

When we examine the list of active compounds of wines and brandies matured in oak, as detailed in this chapter, it is astonishing to note their very great similarity with those compounds present in maple syrup. When will we see the use of maple barrels for aging certain types of wines, or the use of American oak barrels for making maple syrup?

This explains the clear harmonic link between cuisine based on maple syrup and wines matured over a long period of time in oak barrels and casks, such as Sauternes, tawny ports, Madeira wines, and oloroso sherries.

Dry wines, whether white or red, also find their fragrances affected by toasted oak barrels. This allows for a fine pairing with salty cuisine, whose flavor is enhanced by maple fragrances.

Wines raised in American oak barrels are richly endowed with aromatic molecules closely related to those of maple trees, which makes them pair even better with food containing maple syrup. This is the case for New World Chardonnays and some California red wines, as well as Spanish reds from Rioja and Ribera del Duero, where American oak is widely used.

Finally, brandies and whiskies that have matured for quite some time in oak barrels, such as cognac, Armagnac, American bourbon, and Scotch, may also be considered for pairing with some salty dishes, sweet and salty dishes, or dishes sweetened with maple syrup.

## VANILLA/WALNUTS/CARAMEL: PUTTING TOGETHER SIMILAR COMPOUNDS!

The fragrance of walnuts, present in wines raised in an oxidative milieu—as is the case for fino and amontillado sherries and vin jaune from Jura—is a byproduct of the yeast cap that develops at the surface of these wines during their lengthy maturation in barrels, without ullage (in other words, in the absence of contact with the air).

During wine fermentation, the yeasts produce acetaldehydes, compounds that smell like walnuts and are present in sherry and vin jaune (in this last, the acetaldehydes are joined by sotolon) as well as in several wines matured in barrels or bottles for many years.

From a molecular point of view, pure caramel fragrance is very close to that of walnuts.

One aromatic signature of caramel comes from an aldehyde that is closely related to acetaldehyde found in walnuts, as is the case for vanilla. This explains the success of pairing walnuts, caramel, and vanilla in desserts, and of pairing such desserts with wines marked by this type of fragrance. All of these ingredients pair harmoniously with a sweet oloroso or Pedro Ximénez sherry.

# 2. COMPLEMENTARY FOODS

## BARREL-RAISED WINES

APPLES
ASPARAGUS (GRILLED OR ROASTED)
BALSAMIC VINEGAR
BROWN SUGAR
BUTTER
CARAMEL
CARDAMOM
CELERY
CHICORY
CINNAMON
CLOVES
COCOA
COCONUT
COFFEE
CORN SYRUP
CORN TACOS
COTTON CANDY
CURRY
EUCALYPTUS
FOODS COOKED ON A WOOD-FIRED OR CHARCOAL GRILL
FOODS RICH IN UMAMI (SCALLOPS, SEAWEED, CHEESE, CRAB, MEAT, SHIITAKE MUSHROOMS)
GRILLED ALMONDS
GRILLED MEAT
GRILLED SESAME SEEDS
HAVANA TOBACCO LEAVES
HONEY
HAZELNUTS
LEMONGRASS
LICORICE
MADEIRA
MALT POWDER
MANGOS
MEAT (WELL-GRILLED)
MUSHROOMS
PINEAPPLE
POPCORN
ONIONS
OSMANTHUS FLOWERS
PEACHES
QUINCES
ROASTED FENUGREEK SEEDS
SAKE
SANDALWOOD
SCOTCH
SMOKE
SMOKED FISH (THE SKIN OF SALMON COOKED ON ONE SIDE)
SMOKED PEPPERS (MEXICAN CHIPOTLE AND SPANISH PIMENTÓN)
SOY SAUCE
SPICES
STRAWBERRIES
TAMARIND
TEA
TOAST
TONKA BEANS
VANILLA
WALNUTS

ESTERS

TERPENES

# FROM THE STOCKYARD TO THE OVEN

> "Scientific knowledge is the most certain and most useful knowledge that we humans possess."
>
> ALBERT EINSTEIN

In this chapter, we'll take a closer look at beef through the scientific study of its structure and its cooking methods, with the goal of learning how to prepare it using complementary ingredients and to successfully pair it with wines.

It is well known that red meat—such as the very popular beef and lamb—has a richer flavor than does white meat, such as pork. But why is this so? Are there taste differences among different cooking methods for beef, and among different types of beef? What are the complementary ingredients for cooking with these types of meat?

The scientific study of the physical structure of cow's flesh offers some answers to these questions. Such a study can help us choose the meat, its cooking methods, its accompanying ingredients, and the wines that will perfectly accompany it.

Even though we have eaten meat since the days of the Cro-Magnon, we still haven't absorbed all of its gustatory secrets!

Lamb and beef—and the dark parts of so-called white meats (chicken and turkey)—generally contain more substances that generate flavor upon cooking than do the paler types of meat.

## BEEF: A LOOK AT THE ANIMAL

Beef, which is rich in red muscle fibers (slow-twitch fibers, used by the animal for sustained exercise), provides a more flavorful dish than those meats rich in white fibers, which come from animals such as chickens, turkeys, and calves that have not had a great deal of physical exertion.

The animal flavor, characterized by a sensation of fullness (umami) and by rich aromas, comes mainly from meats with red fibers that have undergone exercise. This explains, in part, the choice of wines to accompany red meat; they are generally more aromatic and fleshier, with a greater texture and fuller body than wines selected to accompany white meat.

Meat fat also plays a major role in its flavor. It is the sapid molecules, dissolved in animal fat, that are responsible for a large part of beef's flavor (and for that of lamb, pork, and chicken as well). It's important to know that fat supports flavor. During the cooking process, heated fat bestows on the meat flavors that come from its fruity, floral, hazelnut-like, and vegetal molecules.

Hundreds of new aromatic compounds appear during the chemical reactions that are set off by the heat of cooking meat.

## CATTLE FEED AFFECTS THE COMPLEXITY OF FLAVORS

The type of feed on which cattle are raised also plays an important role in the flavor of the resulting meat.

### BEEF RAISED ON FRESH AND DRIED GRASSES

Red meat from cattle whose diet is based on pasture grass during the summer and hay during the winter is much more flavorful and more aromatic than meat from grain-fed cattle.

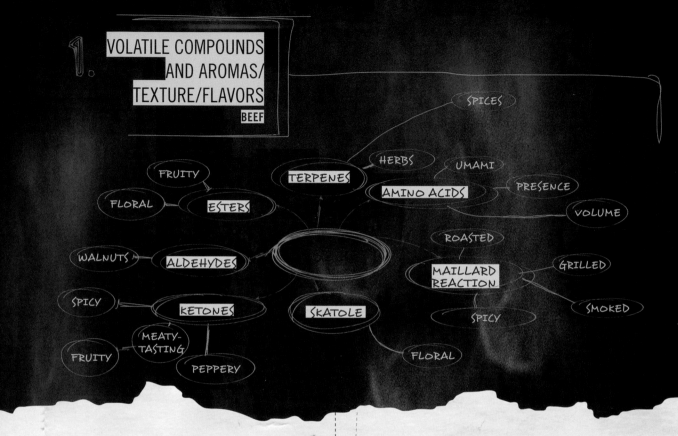

## 1. VOLATILE COMPOUNDS AND AROMAS/TEXTURE/FLAVORS
### BEEF

SPICES

HERBS — TERPENES — UMAMI

AMINO ACIDS — PRESENCE

VOLUME

FRUITY — ESTERS
FLORAL

ROASTED

WALNUTS — ALDEHYDES

MAILLARD REACTION — GRILLED

SPICY — KETONES — SKATOLE

SMOKED

SPICY

FRUITY
MEATY-TASTING
PEPPERY

FLORAL

Grass-fed meat contains a greater quantity of polyunsaturated fatty acids and chlorophyll—which the animals transform into terpenes, a family of aromatic molecules that is strongly present in wine, providing spice and herb notes, among others.

The meat of pasture-raised cattle is also richer in skatole, a molecule that in small quantities imparts a flowery fragrance. In fact, skatole is a volatile compound often used in the perfume industry.

Skatole is part of the indole family, aromatic molecules that smell of orange blossoms, jasmine, beets, and jujubes (*Ziziphus mauritiana*). On the other hand, at an excessive concentration, which (luckily!) edible meat does not have, skatole develops a fecal odor.

The best wines to accompany grass-fed beef are mouth-filling red wines that generate similar flavors (spices, flowers, and herbs); one example is Australian Shiraz.

**Complementary foods for pasture-raised beef:** Based on what we have learned so far, it seems logical that pasture-raised beef should be cooked with spices and herbs. But you may be surprised to learn that it also pairs well with orange blossoms, jasmine, and beets (the juice and the vegetable in any form). These are preparations with which most people aren't familiar.

### GRAIN-FED BEEF

Red meat from grain-fed beef is less tasty than that from grass-fed beef, but on the other hand it is more "beefy" tasting.

Such meat calls for wines that are fairly generous and perhaps a bit rustic to support their taste. Fine choices include

GSM (Grenache/Syrah/Mourvèdre) blends from Languedoc, France, as well as Malbecs, whether from Argentina or Cahors, France.

### AGED BEEF

One must also consider the age of meat, because as it ages, the aromatic molecules contained in the fat develop increasingly powerful flavors. The amino acid level measurably increases during the maturation process, augmenting the beef's presence in the mouth (umami flavor) and thus providing more depth and texture.

This leads us to even more aromatic and voluminous wines that combine well with animal flavors. Fine selections include wines with "hot tannins" from a Mediterranean climate, such as some Cabernet Sauvignons from Australia, California, or Chile, as well as from southern France, and Petite Sirah and Zinfandel wines from California or Mexico.

### ANGUS BEEF

It's a well-known fact: Americans are big beef eaters. This is also true for many Canadians, in particular those from Alberta, where Angus cattle are raised. Angus is widely considered to be one of the best varieties of beef in the world. To satisfy the carnivorous appetite, what could be better than a thick filet of Angus beef surrounded by fried matsutake mushrooms (rich in umami) harvested from the forests of British Columbia?

In contrast to grain-fed beef, the flavor of Angus beef, sprinkled lightly with parsley, with aromas and flavors that have been rendered more complex by the healthy nourishment of the pastures where the cattle are raised, is in the domain of terpenes (aromas of herbs and spices) and skatole (floral fragrances).

Once aged to perfection and cooked properly, Angus beef develops animal notes; roasted, grilled, and smoked tonalities similar to those of wines raised in oak barrels; and notes of onions and walnuts. All of these are caused by the Maillard reaction, a phenomenon that takes place when browning meat at a high temperature.

During maturation and cooking, this meat develops a large quantity of amino acids, resulting in a greater volume, texture, and flavor in the mouth (umami, the fifth basic taste),

compared to other kinds of beef. Such meat should be paired with a full red wine with intense flavors, floral and spicy notes, and woody tonalities that go well with the roasted, grilled, and smoky flavors brought on by cooking.

Which wines to choose? High-quality Pinot Noirs, ideally from the New World, such as the great ones developed by Thomas Bachelder of Clos Jordanne, situated in the heart of the Jordan Bench on Ontario's Niagara Peninsula. One fine example is the dense, complex, extremely rich Le Grand Clos Pinot Noir, which is perched halfway between Burgundy elegance and New World fullness.

### FROM RAW TO COOKED...

Why does raw meat go just as well with a dry medium-bodied white wine as with a light red wine? The answer is simply that raw meat develops relatively few aromas. It is therefore less rich and seems saltier and more acidic in the mouth, two qualities that "break" the overly powerful tannins of full-bodied red wines. So the next time you order steak tartare, don't hesitate to ask for a medium-bodied white wine, but one that's richly perfumed, such as some mature Muscadets that have spent several years in the bottle.

What about cooked meat? The answer is simple: cooking increases its flavor! In fact, the process of browning meat at a high temperature, in particular for grilled, roasted, and braised meat, transforms the raw meat's aromatic molecules into new, more complex molecules.

Beef's naturally occurring esters, ketones and aldehydes, generate aromas of fruits, flowers, walnuts, and vegetables. They attach themselves to the molecules of cooking associated with roasting, grilling, and smoking—similar to the aromas that develop in wine as it matures in toasted oak barrels—as well as those found in spices and even cooked onions.

This explains the lovely, precise union between grilled beef and certain barrel-raised red wines.

### COOKING TECHNIQUES AND WINE

The way in which meat is cooked also greatly influences its aromas and structure, as well as our experience of the wine chosen to accompany it.

## GRILLED AND ROASTED BEEF

Grilled beef and roasted beef develop the greatest number of aromatic compounds from the chemical reactions associated with browning. The meat's original flavors are joined by others, including tonalities of roasting, grilling, smoking, spices, and cooked onion.

With beef prepared in these two ways, you should serve a red wine with a similar aromatic profile, one that is fairly full-bodied and that ideally was raised in oak barrels. There is a wide range of choices—think of wines based on Cabernet Sauvignon, Syrah, Malbec, Tempranillo, Nebbiolo, and Sangiovese. Select the heartiest and most tannic examples.

## BOILED BEEF

The case of boiled beef is quite interesting. First of all, boiling provokes a major change in the meat's structure. In the example of the classic, very French pot-au-feu, when beef is boiled in the cooking liquid, it becomes whiter and, above all, stringier. Its protein fibers bind and lose their "juicy" character, almost in the manner of a white meat.

From a structural point of view, boiled meat has lost its capacity to soften the tannins in red wine. Thus, it is best to serve it with either an easy-to-drink red wine with supple tannins or a rather generous white wine that's moderately acidic. Boiled meat will take on the flavor of its surroundings, so its flavor will depend on the cooking liquid and its ingredients. It usually becomes quite fragrant. Consequently, the selected wine should also be very aromatic.

### SOMMELIER-COOK'S HINT

**"New fragrant water" after cooking** It is recommended that meat that has been cooked in a liquid be chilled in that same cooking liquid. The more a meat is chilled, the greater the capacity of the meat's tissues to retain water. Thus, as it chills, the meat reabsorbs part of the liquid that was lost during cooking.

The molecular gastronomy research of the French physical chemist Hervé This leads us to an interesting suggestion: why not transfer the meat right after cooking into another hot liquid, spiced with your choice of ingredients? As it cools, the meat will expand by about 10% to 15% as it absorbs this "new fragrant water."

Water with truffle oil, with star anise, with licorice, or with tomato? It's up to you to choose your flavor, in accordance with the accompanying wine. You'll be serving meat that may at first seem plain, without any juices, but you'll be subtly pleasing your guests' taste buds—and succeeding with the wine pairing!

### BRAISED BEEF

Generally, braised beef is slightly caramelized before braising, giving it aromatic notes such as those associated with grilled or roasted beef. Its long, slow cooking process on the stove or in a casserole, immersed in an aromatic cooking juice, gives it a structure halfway between juicy grilled or roasted meat and stringier boiled meat. It has a full, almost juicy texture that provides a sensation of richness and fullness, even if the meat can sometimes seem a bit stringy.

This type of meat is made for red wines that are simultaneously strongly fragrant, enveloping, generous, and almost heady, with a velvety thick texture. Many New World wines meet these criteria, in particular GSM (Grenache/Syrah/Mourvèdre) blends and Cabernet-Shiraz, as do wines from regions bathed in the sunshine and heat of the Mediterranean basin, without forgetting Veneto's invigorating Amarones.

### BEEF AT THE TABLE

Certain essential ingredients are well-known for recipes based on beef.

Beef is often served with aromatic herbs and spices, mushrooms, certain cheeses, and tomato in all its forms, including

BEEF. 091

2.

COMPLEMENTARY
WINES
BEEF

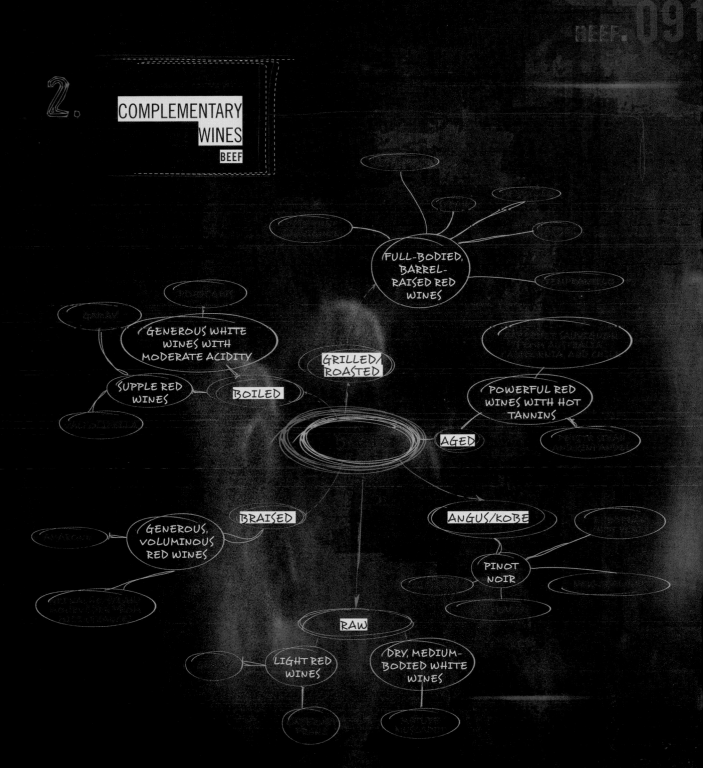

FULL-BODIED,
BARREL-
RAISED RED
WINES

GENEROUS WHITE
WINES WITH
MODERATE ACIDITY

SUPPLE RED
WINES

GRILLED/
ROASTED

BOILED

POWERFUL RED
WINES WITH HOT
TANNINS

AGED

GENEROUS,
VOLUMINOUS
RED WINES

BRAISED

ANGUS/KOBE

PINOT
NOIR

RAW

LIGHT RED
WINES

DRY, MEDIUM-
BODIED WHITE
WINES

## COMPLEMENTARY FOODS
### BEEF

AROMATIC HERBS
ASPARAGUS ROASTED IN
OLIVE OIL
BEETS
CHEESE
COCOA
COFFEE
DARK CHOCOLATE

JASMINE
ORANGE BLOSSOMS
ROASTED NUTS
SAUTÉED MUSHROOMS
SPICES
TOASTED COCONUT
TOMATOES
WILD RICE

## COMPLEMENTARY FOODS RICH IN AMINO ACIDS/UMAMI
### BEEF

AGED CHEESES

ANCHOVIES
BEER
CAVIAR
COOKED ONIONS
COOKED SPINACH
CRAB
DRIED BONITA TUNA
DRIED HAM

DRY SAUSAGE
GREEN PEAS

JAPANESE KOMBU
AND NORI SEAWEED
KETCHUP
MISO
MUSHROOMS

RED TUNA
SCALLOPS
SINGLE MALT SCOTCH
SLOW-COOKED
TOMATO SAUCE
SOY SAUCE

---

that mainstay, ketchup. It is rare, on the other hand, to find beef dishes that include caviar, seaweed, anchovies, or orange blossoms. And yet, all of these ingredients belong to the same molecular sphere as does beef, as they share several aromatic compounds with it.

Among the other ingredients that are complementary to beef and to the wines that accompany it are those influenced by the volatile molecules that appear during cooking. These ingredients include asparagus roasted in olive oil, wild rice, toasted coconut, sautéed mushrooms, cocoa, dark chocolate, coffee, malt, brown and black beer, Scotch, and toasted walnuts.

Finally, we must favor foods rich in amino acids, and thus dominated by the taste of umami, which is produced by the synergy between glutamate (a natural element of certain foods), disodium inosinate, and disodium guanylate. All three of these are natural compounds of umami.

In this last category, we can include the Japanese seaweed kombu and nori, certain mature cheeses (Parmigiano-Reggiano, Emmenthal, and cheddar), dried ham (prosciutto, Spanish ham, and Bayonne ham), red tuna, well-cooked tomatoes, scallops, and mushrooms (shiitake, enoki, and matsutake).

Closely following, with a slightly lower level of umami, are ketchup, soy sauce, caviar, miso, sausages such as chorizo, anchovies, Roquefort and aged Gruyère cheeses, onions, cooked spinach, brown and black beer, and single malt Scotch.

## LAMB: A RED MEAT RICH IN... THYME!

An interesting fact: thymol—the volatile compound responsible for the most important aromatic characteristic of thyme—is also the principal sapid (flavor) molecule contained in lamb. This explains the age-old use of thyme in lamb recipes.

On the other hand, this shakes up the idea of the classic pairing between lamb and red Bordeaux wines, in particular those from Pauillac. So, to keep the pairing from being merely by rote—as is the case with Pauillac—instead serve red wines from the Mediterranean basin, marked by aromas of the garrigue (scrublands where herbs such as thyme are abundant). You can do so even if the lamb recipe contains no thyme. After all, the recipe may not contain thyme, but the lamb does!

## PORK: A WHITE MEAT RICH IN... COCONUT!

Pork's naturally fruity flavor is due to the presence of lactones, a family of active molecules with expressive tones of apricot, peaches, and coconut. This explains why chefs often stuff roast pork with apricots.

This also confirms the successful pairing of pork with white wines whose bouquets belong in the same aromatic sphere. Fine examples include Rhone and California wines based on Roussanne and more exotic wines based on Viognier, as well as certain lightly oaked red wines—the fragrances of the barrel are partially characterized by lactone molecules (see the "Oak and Barrels" chapter)—based on Merlot or Tempranillo. A knowledge of pork's molecular structure also helps explain the regional harmony between aged, dried Spanish ham and a glass of fino sherry, both of which contain lactones (see the "Fino and Oloroso" chapter).

## LACTONES

These volatile molecules occur in more than 120 alimentary products (fruits, vegetables, dairy products, meats, wines, brandies), which explains their importance in the fragrance industry. They are generally responsible for fruity, milky, or caramelized aromatic notes. Aging in toasted oak barrels (see the "Oak and Barrels" chapter) furnishes wines and brandies with a number of lactones.

SKATOLE

GEWÜRZTRAMINER

SCHEUREBE

WOODY

PHENYL
ALCOHOL

PHENYL
ACETATE

PHENYL
ALCOHOL

---

---

# GEWÜRZTRAMINER/GINGER/ LYCHEE/SCHEUREBE

## MOLECULAR FAMILY HISTORIES

> "The more interactions we establish, the better we know the object under study."
>
> JOHN DEWEY

The scientific study of the aromatic structure of foods and wines occasionally helps uncover some surprising molecular family secrets!

For example, when we analyze the aromatic profile of the volatile compounds in lychee and ginger, and of those in white wines made from Gewürztraminer and Scheurebe grapes, we quickly realize their marked similarities.

In order to better understand our four players and fully grasp the interactions among them, I propose a study organized into four themes: lychee and Gewürztraminer (Theme I); Scheurebe and Gewürztraminer (Theme II); Scheurebe and Sauvignon Blanc (Theme III); and ginger, Gewürztraminer, and Scheurebe (Theme IV).

### THEME I
### LYCHEE AND GEWÜRZTRAMINER: TRUE TWINS!

If there is one distinctive aroma in the world of wine, it must be the fragrance of lychee found in Gewürztraminer.

In fact, wines based on this grape variety are so powerfully fragrant that one must, as did Louis de Funès in a scene from his hilarious movie *A Wing or a Leg*, be suffering from ageusia (the loss of taste) if he or she cannot detect lychee's penetrating fragrance, a mixture of volatile compounds with aromatic touches of exotic fruits and flowers, including roses.

Clear similarities have been observed between those volatile compounds that give lychees (fresh or canned) their aroma and the bouquet of white wines based on Gewürztraminer. These similarities make lychees and Gewürztraminer almost identical twins, as if born from a single egg.

For their part, Gewürztraminer and the Austrian Scheurebe grape are also twins, but in this case fraternal rather than identical (see Theme II).

### THE MOLECULAR STRUCTURE OF LYCHEE AND GEWÜRZTRAMINER

Lychee and Gewürztraminer share a good dozen volatile compounds: phenyl acetate (chocolate, flowers, wildflower honey), phenyl alcohol (cocoa, flowers), beta-damascenone (grapefruit, tequila), ethyl butanoate, cis-rose oxide, citronellol, furaneol, geraniol, ethyl hexanoate/ethyl isohexanoate, ethyl isobutyrate, linalool, and vanillin.

The cis-rose oxide (which has a powerful scent of roses and lychee) is the strongest aromatic signature of both lychee and Gewürztraminer.

This is followed very closely by linalool (a terpene with the floral scent of lavender and lilies of the valley), geraniol (whose complex fragrance is reminiscent of roses/germaniums), and citronella/lemon.

There is also beta-damascenone (with touches of exotic fruits and roses), furaneol (with the fragrance of caramelized strawberries/pineapple and cotton candy/caramel), phenyl alcohol (smelling of rose and peaches, with touches of

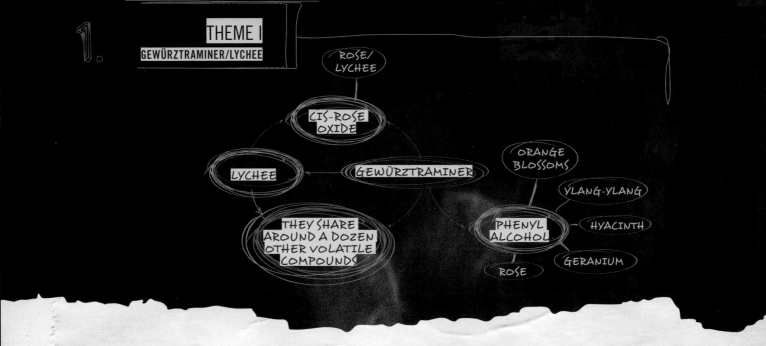

ROSE/LYCHEE

CIS-ROSE OXIDE

LYCHEE

GEWÜRZTRAMINER

ORANGE BLOSSOMS

YLANG-YLANG

THEY SHARE AROUND A DOZEN OTHER VOLATILE COMPOUNDS

PHENYL ALCOHOL

HYACINTH

GERANIUM

ROSE

hyacinth/orange blossoms/ylang-ylang/geranium), and ethyl hexanoate (smelling of pineapple and bananas).

If several different molecules (from a structural point of view) remind us of fragrances associated with the same foods or complementary ingredients, it's because these foods and ingredients contain those molecules and not because a single molecule has several fragrances.

All of these elements combine to create the penetrating fragrance found in both lychees and Gewürztraminer.

In Gewürztraminer, the rose/lychee aroma generated by the cis-rose oxide is supported by linalool; geraniol, phenyl acetate, and phenyl alcohol, all four of which are volatile compounds with tones that are also quite floral. These compounds reinforce Gewürztraminer's floral aspects and its harmonious power of attraction for recipes dominated by ingredients with these same floral properties.

## CINNAMON AND GEWÜRZTRAMINER

Late-harvest Gewürztraminer-based wines are often marked by fragrances that come from cinnamic aldehyde (cinnamon), eugenol (cloves), linalool (flowers), cineol (eucalyptus), and camphor, all compounds that form a strong part of cinnamon's identity. This is the reason for the well-known pairing of this spice and Gewürztraminer wines (see the "Cinnamon" chapter).

## LYCHEES AND GEWÜRZTRAMINER AT THE TABLE

Given the forceful molecular power of attraction between these two entities, dare to give lychees a starring role in your cuisine; you will see that it pays off greatly when it comes to pairing your meal with Gewürztraminer-based wines.

Here are some ideas for lychee-based dishes—some savory, some sweet—that work well with dry or late-harvest Gewürztraminer wines:

+ Chicken with hot peppers and lychees; **Spiced lychee tartare** (see recipe in *À table avec François Chartier*); crystallized ginger cotton candy served on a lychee soup; gingered lychees with citrus salsa; lychee panna cotta with lime zest; pineapple salad with lychees perfumed with rosemary.

+ Let's not forget **Lychee granita with creamy white chocolate yogurt, pink grapefruit, Campari jelly, and hibiscus flowers**, conceived and realized by the dessert maestro Patrice Demers (see recipe on page 103). A multi-layered symphony of flavors, this recipe could have been designed expressly for a late-harvest Gewürztraminer such as the great minerally Gewürztraminer Clos Windsbuhl Vendanges Tardives 2005, Domaine Zind Humbrecht, Alsace.

For a true understanding of the potent attraction between lychee and Gewürztraminer, try pairing an entree of Asian-style chicken with hot peppers and lychees and a dry, forceful Alsatian wine such as Gewürztraminer Wintzenheim 2005, Domaine Zind Humbrecht, Alsace. This wine is simultaneously powerful, minerally, and very spicy, with an ample, full texture, expressing itself through lychee, pineapple, and rose flavors. Its dense texture serves only to bring out the peppers' fire, while implementing the attraction between its aromatic compounds and those of lychee.

THEME II
SCHEUREBE AND GEWÜRZTRAMINER:
FRATERNAL TWINS

The Austrian grape variety Scheurebe (also known as Sämling 88) was created through the botanical crossbreeding of Riesling and an unknown grape variety. For a long time, it was believed that this unknown parent was Silvaner, but a recent study of its DNA rejected this notion, although it did confirm Riesling as a parent.

This crossbreeding, performed in Germany in 1916 with the goal of creating a more aromatic grape variety, is what led me to suspect the molecular rapprochement between Scheurebe and Gewürztraminer. In fact, the latter may very well prove to be the second, unknown parent of Scheurebe...

Scheurebe nevertheless remains, from a structural point of view, in particular at the level of its most volatile compounds, a twin of the Gewürztraminer grape variety. We could call it a fraternal twin—a false twin born from a second egg—in contrast to identical twins, born from a single egg, which seems to be the case for lychee and Gewürztraminer.

To understand the aromatic link between Scheurebe and Gewürztraminer, you need only have enjoyed the late Alois Kracher's remarkable, highly fragrant Scheurebe-based Austrian sweet wines. These occupy the same exalted territory as the finest noble rot *(botrytis cinerea)* wines of the world, such as Chateau d'Yquem Sauternes and the offerings of Egon Müller in Mosel, Germany, and István Szepsy in Tokaj, Hungary.

There is no doubt about it: there is a compelling connection between the aromas detected in the two following Austrian Scheurebe wines and those of certain Alsatian Gewürztraminers:

+ **Scheurebe TBA No. 4 "Zwischen den Seen" 2001 Burgenland, Kracher, Austria** This Scheurebe wine, raised in stainless steel for eighteen months, contains 203.6 grams of residual sugar per liter and 11% alcohol. It has a color of old Italian gold with orange tints and a nose that's simultaneously subtle and explosive, with intriguing aromas of smoke, strawberry/pineapple, passion fruit, and lychee—it couldn't be more Gewürztraminer! This masterpiece has a mouthfeel that is nervous, lean, and straightforward, with a magnificent smoothness balanced by a cleansing acidity, and it brings to mind the finest Alsatian Sélection de Grains Nobles (SGN) noble grape wines.

This wine, a personal favorite of mine, was served during the media launch of *La Sélection Chartier 2006* at the former Quebec City restaurant L'Utopie, accompanied by a dish created by the acrobatic master chef Stéphane Modat: a Roquefort panna cotta, three fruits in a tangy crust, with aged Grenache caramel and topped with sprigs in three flavors. The wine created a vibrant harmony with this "cheese-dessert" (blue cheeses share some of the chemical compounds found in Gewürztraminer and Scheurebe). And it was a great discovery for many of the invited guests!

+ **Scheurebe TBA No. 9 "Zwischen den Seen" 2001 Burgenland, Kracher, Austria** This wine is made from individually selected, noble Scheurebe grapes. Raised for eighteen months in stainless steel vats, it spent more than two years completing its fermentation, finishing the process with a mere 5.5% alcohol. Despite its 317.0 grams of residual sugar per liter, this ultra-preserved, sweet wine presents an unusual level of crispness for a white wine made from such ripe grapes. Its golden color is followed by the powerful fragrance of crystallized ginger, lychee, and white pepper, as in fine Alsatian Gewürztraminers. In the mouth, its sugar unrolls like a flying carpet, in a surprisingly airy and forward-thrusting style. The genius of the late Alois Kracher lay in his ability to find the perfect point of equilibrium between the sugar and the acidity, which he did with the skill of a tightrope walker. I tip my hat to him!

Now imagine this wine served with a spiced lychee tartare, for which you can find a recipe in *À table avec François Chartier*. Spices and lychees: two fine companions that bring out the best in a truly great wine.

## A SUBTLE AROMATIC DIFFERENTIATION...

Why aren't Gewürztraminer and Scheurebe identical twins, as are lychee and Gewürztraminer? The answer can be found by uncovering the two aromatic signatures that characterize each of these two grape varieties.

In Gewürztraminer, the aromatic molecule cis-rose oxide, with the powerful aroma of roses and lychee, is its trademark; this compound is absent from Scheurebe-based wines.

Scheurebe stakes out its identity with 4-mercapto-4-methylpentan-2-one (a molecule absent from Gewürztraminer wines), a sulfuric compound belonging to the thiol family, with a complex aroma reminiscent of boxwood, black currants, pink grapefruit, and passion fruit (see Theme III for more details on this volatile compound).

It is important to note that 4-mercapto-4-methylpentan-2-one, which is present in some Scheurebe-based wines, was discovered by the research team of Bordeaux wine scientist and professor Denis Dubourdieu while researching the Sauvignon Blanc grape—which explains a certain link between these two grape varieties (see Theme III).

The majority of the other aromatic compounds found in our two fraternal twins are present in wines made from both grape varieties. Such is not the case for cis-rose oxide, which is unique to Gewürztraminer, nor for 4-mercapto-4-methylpentan-2-one, which is linked to Scheurebe.

## THE MOLECULAR STRUCTURE OF GEWÜRZTRAMINER AND SCHEUREBE

Following are the aromatic compounds—all providing fruity/floral as well as sugary/woody tones—that these two grape varieties share and that make them, from a molecular point of view, fraternal twins:

+ **Isoamyl acetate** (banana aroma, also present in ripe apples)
+ **Beta-damascenone** (carotenoids with a fruity, floral smell): The smell of exotic fruits (passion fruit, mango, kiwi, carambola), as well as rose, tobacco, and black tea. These are also found in apricots, raspberries, blackberries, rum, and wine. During its aging process, beer develops floral, fruity tones that come from the formation of beta-damascenone.

+ **Ethyl butanoate** (an ester with the fragrance of pineapples, apples, and tutti frutti, also used to artificially flavor some alcohols and cocktails, such as martinis and daiquiris)
+ **Ethyl hexanoate** (sweet, waxy odor, pineapple, green bananas)
+ **Ethyl isobutyrate** (sugary, fruity, etherlike odor, with touches of rum and eggnog)
+ **Linalool** (a terpenic alcohol smelling of lavender and lilies of the valley, also the principal compound in bergamot, rosewood, mint, citrus fruits, and cinnamon)
+ **Ethyl octanoate** (fruity, waxy aroma associated with bananas, apricots, pears, and brandy)
+ **Wine lactones** (intense sugary aromas and woody aromas, associated with coconut, apricots, and peaches)

## THEME III
## SCHEUREBE: A CLOSE COUSIN OF SAUVIGNON BLANC?

Depending on the degree to which 4-mercapto-4-methylpentan-2-one expresses itself—in the aromatic realm of boxwood, black currants, pink grapefruit, and passion fruit—Scheurebe can either be considered a cousin of Sauvignon Blanc or a fraternal twin of Gewürztraminer.

When Scheurebe's other aromatic compounds dominate 4-mercapto-4-methylpentan-2-one, the wine expresses more floral/lychee notes, which, as we previously discussed, make it the fraternal twin of Gewürztraminer. We can thus rely on wine and food pairings appropriate to Gewürztraminer for Scheurebe wines with a more floral/lychee profile.

Scheurebe, however, becomes a close cousin of Sauvignon Blanc if 4-mercapto-4-methylpentan-2-one dominates the other components, providing notes linked with Sauvignon Blanc, such as boxwood, red and black currants, grapefruit, and passion fruit.

We also find in Sauvignon Blanc 3-mercaptohexan-1-ol and 3-mercaptohexyl acetate, members of the same family of compounds found in varying proportions in Gewürztraminer, Riesling, Colombard, and Petit Manseng wines. The compound 3-mercaptohexan-1-ol is found in passion fruit, while 3-mercaptohexyl acetate brings passion fruit and boxwood to mind.

Scheurebe wines whose aromatic profile is associated with 4-mercapto-4-methylpentan-2-one (again, with aromas of boxwood, black and red currants, pink grapefruit, and passion fruit), are therefore best paired with those dishes usually suggested for late-harvest Sauvignon Blanc wines or Sauternes and similar wines, selecting blends containing a large portion of Sauvignon Blanc grapes.

## THEME IV
## GINGER, GEWÜRZTRAMINER, AND SCHEUREBE:
## A HEAVENLY TRIO

Ginger (see the chapter by that name) contains several volatile compounds that endow it with floral, citrus, woody, and camphor-like tones, in the aromatic image of Gewürztraminer and its fraternal twin, Scheurebe (at least, those Scheurebes of the lychee/rose variety).

These three entities all share the following dominant aromatic compounds: linalool, geraniol, camphene, neral, and limonene.

Given this strong force of molecular attraction, Gewürztraminer and Scheurebe are the grape varieties that link most harmoniously with foods dominated by ginger and its complementary ingredients, which include turmeric, cardamom, and galangal (for other complementary ingredients, see the "Ginger" chapter).

Imagine the magnificent union that happens when ginger and its complementary ingredients are accompanied by lychee. We attain a nirvana of familial harmony!

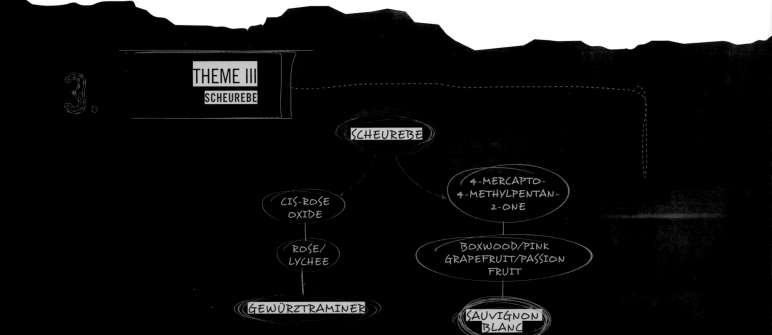

THEME III
SCHEUREBE

3.

SCHEUREBE

CIS-ROSE OXIDE

4-MERCAPTO-4-METHYLPENTAN-2-ONE

ROSE/LYCHEE

BOXWOOD/PINK GRAPEFRUIT/PASSION FRUIT

GEWÜRZTRAMINER

SAUVIGNON BLANC

**SUMMARY**

## IDENTICAL TWINS

LYCHEE/
GEWÜRZTRAMINER

CIS-ROSE
OXIDE

ROSE/
LYCHEE

ROSE

PHENYL
ALCOHOL

GERANIMUM

ORANGE
BLOSSOM

MORE THAN
A DOZEN
COMPOUNDS

## FRATERNAL TWINS

SCHEUREBE/
GEWÜRZTRAMINER

SAME COMPOUNDS
WITH TWO
EXCEPTIONS

4-MERCAPTO-
4-METHYLPEN-
TAN-2-ONE
(SCHEUREBE)

CIS-ROSE OXIDE
(GEWÜRZTRAMINER)

## MOLECULAR TRIPLETS

GINGER/
SCHEUREBE/
GEWÜRZTRAMINER

SHARE SEVERAL
CLOSELY RELATED
VOLATILE
COMPOUNDS

## SOMETIMES COUSINS...

SCHEUREBE/
SAUVIGNON
BLANC

4-MERCAPTO-
4-METHYLPENTAN-
2-ONE

SAME COMPOUND
SOMETIMES
DOMINANT

LYCHEE GRANITA WITH CREAMY WHITE CHOCOLATE YOGURT,
PINK GRAPEFRUIT, CAMPARI JELLY, AND HIBISCUS FLOWERS

RECIPE CONCEIVED AND REALIZED BY PATRICE DEMERS,
MASTER PASTRY CHEF

TASTE BUDS AND MOLECULES

## LYCHEE GRANITA WITH CREAMY WHITE CHOCOLATE YOGURT, PINK GRAPEFRUIT, CAMPARI JELLY, AND HIBISCUS FLOWERS

### CAMPARI JELLY

+ 200 g grapefruit juice (note: pink is best)
+ 60 g sugar
+ 3 packages of rehydrated gelatin
+ 75 g Campari

In a small pan, bring to a boil the grapefruit juice and sugar. Remove from the heat, skim off the impurities, and add the gelatin and the Campari. Pour into a small, lightly oiled container and cover with plastic wrap. Let chill for at least 8 hours, then cut the jelly into small cubes.

### CRUNCHY HIBISCUS

+ 350 g peeled, cleaned green apples, cut into cubes
+ 150 g water
+ 100 g isomalt
+ 3 tbsp dried hibiscus
+ 1 egg white

In a small pan, cook the apples, water, isomalt, and hibiscus over a medium flame until the apples are quite tender. Blend to a smooth purée and pass through a cheesecloth. Bring to room temperature and add the egg white. Place on a silicon mat and cook at 110°C (230°F) for about an hour until the purée is completely dry, at which point it will break into chips. Store in a sealed container.

### GRAPEFRUIT

+ 1 pink grapefruit (note: pink grapefruits are best for pairing with Gewürztraminer wine)
+ 100 g sugar
+ 200 g water

Peel the grapefruit and cut into segments (supremes). Cut each supreme in three. Make a syrup from the sugar and water and pour it onto the grapefruit. Store in the refrigerator.

### CREAMY WHITE CHOCOLATE YOGURT

+ 150 g cream
+ 240 g white chocolate
+ 215 g Mediterranean yogurt

In a pan, bring the cream to a boil and pour it over the chocolate in a pot. Let stand for one minute and then emulsify with a whisk. When the chocolate is well mixed, add the yogurt, mix well, and let chill for at least 8 hours.

### LYCHEE GRANITA

+ 250 g lychee purée
+ 25 g sugar
+ 5 g lemon juice

Heat the sugar with one third of the purée, remove from the heat, and add the remaining purée and lemon juice. Let the mixture thicken in the freezer.

### ASSEMBLY

Garnish bowls with the grapefruit segments and the jelly cubes. Using a pastry bag, cover this with a layer of the creamy yogurt. Add the lychee granita and garnish with the hibiscus chips, edible flower petals (note: ideally rose, lavender, geranium, or osmanthus flowers, preparing for an eventual pairing with Gewürztraminer), and wild rose powder. This can be made from crushing dried rose petals, which you can find in specialty stores or online at www.epicesdecru.com.

Dare to play around with this recipe, which to my mind is a good partner for a fine sweet Alsatian Gewürztraminer, as well as certain Austrian Scheurebe-based wines. For example, try adding or interchanging ginger, lemongrass, rosewater, orange blossoms, lavender, osmanthus flowers, or the molecular pair of pineapple and strawberries.

FRUITY
ESTERS

EUGENOL

SAUTERNES AND ITS
NEIGHBORS
FRUITY ESTERS
FURANEOL
EUGENOL
ETHYL BUTANOATE
WINE-TYPE
FRAGRANCE

# PINEAPPLE AND STRAWBERRIES

## A STRANGE OVERLAPPING DESTINY

"New ideas are always suspect and generally resisted,
for the sole reason that they aren't yet widespread."

JOHN LOCKE

Research in molecular biology has shown that pineapple and strawberries, despite their major differences in color and structure, share several volatile compounds that express themselves through numerous identical aromatic molecules.

This explains the strange aromatic familiarity that one would experience if one were to taste pineapple and strawberries one after the other wearing a blindfold. This is even more the case if both are very ripe or cooked.

Let's take a closer look at the overlapping destiny of these "molecular twins," as well as the wines that work in perfect harmony with this inseparable duo.

### PINEAPPLE

The taste of pineapple (*Ananas sativus*)—whose Latin name comes from *nana*, an American Indian word for "perfume"—is intensely sugary and often has a sharply acidic finish. It is composed of multiple aromatic compounds.

Among the dominant compounds found in pineapple are eugenol, with an aroma of cloves; vanillin, with an aroma of vanilla; fruity esters smelling of pineapple and basil; oxygenated compounds of the carbon cycle such as furaneol, which expresses caramel-like tonalities; and other compounds whose aromatic profiles are sherry-like, bringing to mind amontillado and oloroso sherries.

When the fruit is overripe, there is in addition to these dominant aromatic compounds an almost animal touch, with a meaty profile. As with other fruits, pineapple's fragrance is dominated by esters such as ethyl butanoate, also known as butyrate.

We also find pentanone, smelling of wine and acetone, which is also present in barrel-raised brandies, bananas, citrus fruits, and grapes. Finally, pineapples contain ethyl propanoate, a fruity ester with a strong pineapple/strawberry fragrance.

Pineapples share with strawberries, as you will see later in this chapter, those volatile compounds responsible for different aromatic notes: furaneol (caramel fragrance), eugenol (clove fragrance), and certain fruity esters (pineapple/strawberry fragrance).

It is interesting to note that because the volatile compounds associated with pineapple and strawberries (as well as with vanilla and rosemary) contain a healthy portion of eugenol, these two fruits also have a close relationship with cloves and with those wines that pair well with them. The touch of cloves (see the "Cloves" chapter) is particularly noticeable when the pineapple and strawberries are very ripe.

A homemade dish that features pineapple, strawberries, rosemary, vanilla, or cloves should make an interesting pairing with eugenol-rich wines that are recommended for dishes dominated by cloves and vanillin—today, vanillin is synthesized from eugenol derived from cloves.

The strange overlapping destiny of pineapple and strawberries, which has made them almost molecular twins, gives

## VOLATILE COMPOUNDS
### PINEAPPLE AND STRAWBERRIES

FRAGRANCE
FRAGRANCE
ETHYL PROPANOATE
PENTANONE
VANILLIN

**PINEAPPLE**

FRUITY ESTERS
FURANEOL
EUGENOL
ETHYL BUTANOATE
-TYPE
FRAGRANCE

**STRAWBERRIES**

ETHYL CINNAMATE
PENTENAL
FRAGRANCE
VEGETAL AND GREEN
LEAF FRAGRANCES

---

us a new tuning fork with which to create harmony, both among the dishes that feature these foods and with the wines that accompany them.

### PINEAPPLE: A MEAT TENDERIZER

Pineapple—just like papaya, melon, kiwi, and fresh ginger—contains a particular enzyme, bromelain, that has the power to tenderize meat by breaking down the gelatin that it contains. To use pineapple (and the other ingredients named here) in desserts or in savory dishes that contain gelatin, it's necessary to cook the pineapple first in order to neutralize these enzymes. Otherwise, the gelatin won't hold.

### PINEAPPLE (AND/OR STRAWBERRIES) WITH WINE

Among the most important compounds responsible for the pineapple aroma in wines are isoamyl butanoate (apple aroma) and ethyl butanoate (pineapple aroma), two esters that intermingle to create the "pineapple" profile.

Sauternes and its neighbors (Cadillac, Loupiac, and Sainte-Croix-du-Mont) give off an intense fragrance in which pineapple can be omnipresent—as can vanilla and cloves, both of which are endowed with the same aromatic compounds as pineapple.

This explains the harmony between these sweet wines and dishes that are sweet, savory, or a mixture of the two, such as a duck cutlet accompanied by duxelles of pineapple caramelized with vanilla or cloves. The same goes for desserts made from pineapple and enhanced with strawberry, vanilla, or cloves.

Among dry whites, Chardonnay is the grape variety most richly endowed with those exotic notes that bring pineapple to mind. Cool-climate Chardonnays, such as those from Chablis, generate a fragrance of very cool, underripe pineapple, whereas warm-climate Chardonnays, such as those from Australia, California, and Chile, remind one of very ripe pineapple or canned pineapple in syrup.

Other dry white wines, such as creamy Jurançon Sec and Pacherenc du Vic-Bilh Sec, largely based on Petit Manseng and Gros Manseng grape varieties, also pair very well with pineapple and strawberries. These wines nearly constantly exhale complex tones associated with several aromatic compounds of pineapple and ripe strawberry, as well as of their complementary ingredients, vanilla and cloves.

White wines with curry fragrances, such as those from Jura, France, whether Chardonnay or Savagnin, and those from the Rhone Valley, based on Marsanne and Roussanne, also work very well with pineapple-based dishes.

### PINEAPPLE AND CURRY

There is a natural link between pineapple and curry, which explains the presence of pineapple in certain Indian curries. During cooking, pineapple develops, especially when caramelized, volatile molecules from the sotolon family. Other members of this family include curry, roasted fenugreek seeds, soy sauce, and sweet wines made from grapes subjected to noble rot (*botrytis cinerea*), such as Sauternes (see the "Sotolon" chapter). That's the explanation!

Finally, given that very ripe pineapple is rich in oxygenated carbon-containing compounds such as furaneol, which expresses itself through tones of caramel and sherry (amontillado and oloroso), one can create some very interesting pairings between these sherries and desserts that combine caramel and pineapple, such as a pineapple or strawberry tarte tatin, spiced with cloves or vanilla or both.

In the world of savory/sweet dishes, the traditional ham with pineapple, which is usually glazed with brown sugar or maple syrup and is therefore rich in caramelized tones, works very well with a bracing sherry, an oloroso-style Montilla-Moriles, or a creamy Jurançon. And you'll get the same success if you replace the ham's pineapple with strawberries!

When strawberries and pineapple are part of a dish that calls for a red wine, above all choose Spanish vintages (raised in oak barrels, and so stamped with eugenol and vanillin) based on Grenache (Priorat, Campo de Borja, Cariñena), a Tempranillo/Grenache blend (Rioja Baja), or a Mencia (Bierzo).

You can also count on red wines based on Touriga Nacional (Portugal), New World Pinot Noir (California, New Zealand), Zinfandel (California), or Petite Sirah (Mexico).

### REPLACE PINEAPPLE WITH STRAWBERRIES?

Given that pineapple shares numerous aromatic molecules with strawberries (you'll read more about this in the following section on strawberries), making these two fruits almost molecular twins, don't forget that if in your recipes you replace pineapple with strawberries, you should select the same wines.

### STRAWBERRIES

The strawberry variety most often cultivated in North America is a hybrid whose scientific name is *Fragaria x ananassa*. *Ananassa* comes from *ananas*, the Latin word for pineapple. Those who bred this variety at the end of the eighteenth century correctly noticed that its fragrances reminded them of pineapple.

The majority of strawberries cultivated today come from this European hybrid, based on two American species accidentally created in Great Britain in 1750. The aromatic touch of pineapple usually found in ripe strawberries comes from fruity esters, sulphur-containing compounds, and an oxygenated molecule, furaneol.

The molecule responsible for the aroma of very ripe strawberries is in fact the aforementioned furaneol, an aromatic compound smelling of caramel that is also found in ripe pineapple. The aromatic composition of strawberries, like that of pineapple, is characterized by notes of green leaves, caramel, pineapple, cloves, and Concord grapes.

As I mentioned previously, strawberries and pineapple share the volatile molecules responsible for certain aromatic touches: furaneol (caramel), eugenol (cloves), and some fruity esters (pineapple/strawberries).

Strawberries also contain the following volatile compounds: pentenal (with the vegetal fragrance of apples/oranges/strawberries/tomatoes; also present in cognac), ethyl butanoate (also called ethyl butyrate, with the fragrance of pineapple), methyl cinnamate, and ethyl cinnamate.

These last two are esters of cinnamic acid. Ethyl cinnamate is also present in cinnamon, and has a cinnamon/balsamic/honey fragrance and the sugary taste of apricots and peaches. Methyl cinnamate is also present in galangal, Sichuan pepper, certain varieties of basil, and *Eucalyptus olida*.

*Eucalyptus olida*, a tree native to Australia, also carries the evocative name of strawberry gum. It contains a very high percentage of methyl cinnamate, which is used to reproduce the aromas of strawberries and cinnamon in the food and perfume industries.

## 2. COMPLEMENTARY FOODS
### PINEAPPLE AND STRAWBERRIES

AMONTILLADO AND OLOROSO
SHERRIES
APPLES
APRICOTS
BALSAMIC VINEGAR
BANANAS
BARREL-RAISED BRANDIES

BASIL
CARAMEL
CINNAMON

CITRUS FRUITS
CLOVES
CONCORD GRAPES
CURRY
EUCALYPTUS
GALANGAL
GINGER
GUINEA PEPPER
HONEY
MAPLE SYRUP
PEACHES

PINEAPPLE AND/OR
STRAWBERRIES
ROASTED FENUGREEK SEEDS
ROSEMARY
SHERRY
SICHUAN PEPPER
SOY SAUCE
TOMATOES
TURMERIC
VANILLA

## 3. COMPLEMENTARY WINES
### PINEAPPLE AND STRAWBERRIES

PINEAPPLE/
STRAWBERRY

CADILLAC

SAINTE-CROIX-
DU-MONT

SAUTERNES AND
ITS NEIGHBORS

LOUPIAC

CHABLIS

ZINFANDEL

PETITE SIRAH

RED BARREL-
RAISED WINES

JURANÇON DRY/
CREAMY

JURA

NEW WORLD

PINOT NOIR

CHARDONNAY

NEW WORLD

BIERZO

MENCIA

PACHERENC DU VIC-BILH
DRY/CREAMY

GRENACHE
(GARNACHA)

RIOJA

PRIORAT

SAVAGNIN

SHERRY

ROUSSANNE/
MARSANNE

RHÔNE

MONTSANT

JURA

OLOROSO

AMONTILLADO

AUSTRALIA

LANGUEDOC

STRAWBERRY-GLAZED HAM

## INTERCHANGING PINEAPPLE AND STRAWBERRIES

In conclusion, since pineapple and strawberries share a large percentage of their aromatic molecules, it is easy to replace pineapple with strawberries in recipes and to successfully pair strawberry dishes with wines normally recommended to accompany pineapple.

Have some fun in the kitchen by preparing a ham glazed with strawberries (instead of pineapple), a pork sauté with strawberries, or a strawberry tarte tatin. The substitution can also be done in reverse for dishes normally made with strawberries; for example, a strawberry and rhubarb tart can become a pineapple and rhubarb tart. Dare to play around with these substitutions, while still respecting the same wine pairings. You might come out with some unusual treats, such as a pineapple shortcake with rosemary-flavored whipped cream!

## SOMMELIER-COOK'S HINT

**Pineapple scented with fennel and star anise, Ferran Adrià–style** Impress your guests with ideas perfected in Ferran Adrià's famous workshop and published in the remarkable and very accessible book *A Day at elBulli* (Phaidon). These ideas can actually be quite easy to execute; this recipe—which stems from Adrià's work on infusion methods without liquid—serves as proof. It allows the pineapple's aromatic molecules to intermingle with the volatile compounds of the family of anise-flavored foods, which are found in Sauvignon Blanc and Petit Manseng wines, as well as in fennel and star anise. In short, the ingredients and the wine seem made for each other.

In an airtight container, place a layer of fresh fennel fronds and a few pieces of Chinese star anise. Place fresh pineapple segments on top. Cover with another layer of fresh fennel fronds and Chinese star anise. Close the container and let the mixture sit for seven hours. As I have said throughout this chapter, don't hesitate to replace the pineapple with strawberries! Serve the dish right in this "fennel nest" and experience the harmony on your taste buds when you pair it with an exotic, spicy, mellow, and subtly anised Jurançon, such as the superb Symphonie de Novembre Jurançon 2004, Domaine de Cauhapé, France.

APRICOTS
BEER
CHINESE ROSEBUDS
CINNAMON
COFFEE
COOKED ASPARAGUS
FIVE SPICES
FOUR SPICES
GINGERBREAD
GRILLED BEEF
HAZE
IND

TOA
TOASTE
VANILLA
WILD BASIL

CLOVES

ACETYLEUGENOL

WARM AND
SWEET ODORS

# CLOVES

## THE SPICE OF THE BARREL

"All science is certain and evident knowledge."

RENÉ DESCARTES

At the table, there are certain ingredients whose motto could truly be "Open, Sesame." These ingredients lead the way toward food and wine pairing nirvana. Such is the case for cloves.

Following are some new ideas for tapping into the harmonic potential of the remarkable clove, including a look at its complementary ingredients and the wines that pair perfectly with it (as you will see, it has an astonishing link to wines raised in oak barrels).

### THE ORIGIN AND MOLECULAR COMPOSITION OF CLOVES

Cloves are the flower buds of an evergreen tree from the Myrtaceae family, whose members include eucalyptus, myrtle, and the guava tree. Clove trees are cultivated mostly on the island of Zanzibar and neighboring Pemba, which is by far the largest producer of cloves. Following with a much lower production rate are Guadeloupe, Indonesia (the largest consumer), Madagascar, Martinique, Mauritius, Réunion Island, and the Seychelles Islands.

### THE MOLECULAR STRUCTURE OF CLOVES

Cloves are the spice with the highest concentration of aromatic molecules. Very fresh, high-quality cloves may contain up to 20% of their molecular weight in essential oils. That's enormous! Moreover, this high concentration explains cloves' powerful, persistent flavor.

Cloves contain from 70% to 90% eugenol (the principal fragrance of cloves) and from 3% to 12% caryophyllene (which has a woody fragrance).

They also contain from 2% to 3% acetyl eugenol (a hot, sweet fragrance) and 2% oleanolic acid (a triterpenoid smelling of fir trees), as well as some traces of vanillin (vanilla fragrance) and furfurals (found in sawdust and maple syrup, and attaining significant levels in wines aged in oak casks). The penetrating odors of furfurals play a role in the complex tonalities that are simultaneously sugary, woody, and smelling of caramel and hazelnuts, with nuances of toast, burnt food, and coffee.

Eugenol is also the dominant volatile compound in Thai basil (70% of its volatile content), wild basil (60%), malted barley (thus also beer and Scotch), apricots, pineapples, cooked asparagus, roast beef, cinnamon, strawberries, fresh mango, mozzarella cheese, potatoes, and rosemary.

### THE OAK BARREL

As we saw in the "Oak and Barrels" chapter, eugenol, the dominant volatile compound in cloves, is one of the principal—and, to my mind, one of the most characteristic—aromatic signatures generated by charred oak barrels. As was mentioned in that chapter, eugenol is particularly dominant in wines raised in oak from Russia or from the Centre region of France. Even though American oak barrels produce a lower level of

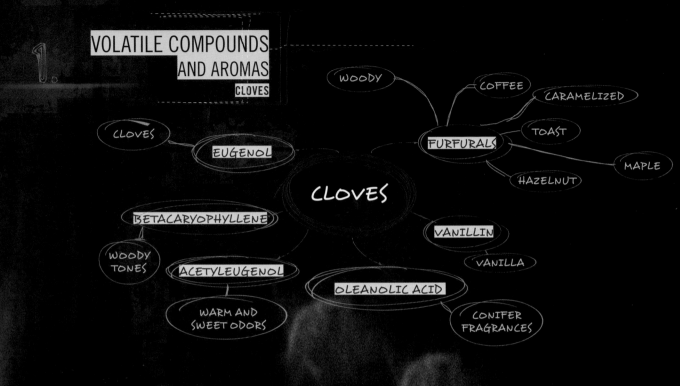

CLOVES — EUGENOL — CLOVES

FURFURALS — WOODY, COFFEE, CARAMELIZED, TOAST, MAPLE, HAZELNUT

BETACARYOPHYLLENE — WOODY TONES

ACETYLEUGENOL — WARM AND SWEET ODORS

VANILLIN — VANILLA

OLEANOLIC ACID — CONIFER FRAGRANCES

eugenol, wines raised in them also show strong aromatic traces of cloves.

Because cloves also contain traces of aromatic compounds such as furfural (a volatile compound found in oak barrels and sawdust, among other places), caryophyllene (a molecule with a woody fragrance), and vanillin (one of the principal aromatic compounds of oak barrels), the aromatic link between cloves and wines raised in oak barrels is thus reinforced.

We also saw in the "Oak and Barrels" chapter that the very substantial, sensual fragrance of cloves makes them a good match for warm, penetrating red wines, especially certain eugenol-rich Spanish vintages. The best wine companions for cloves are Spanish reds based on Garnacha (Grenache) grapes from the Priorat, Rioja Baja, Campo de Borja, and Cariñena regions; wines based on Tempranillo grapes from the Rioja, Toro, and Ribera del Duero regions; and wines based on the Mencia grape, cultivated in Bierzo.

Other fine candidates are Portuguese wines based on Touriga Nacional, a variety from the Douro region, as well as many Pinot Noirs, whether from Burgundy or New Zealand. The barrel's origin is of little importance. Certain grape varieties, such as those mentioned above, often seem to be marked by the aromatic signature of cloves, which probably comes from the lignins in the grape stems.

The same phenomenon occurs with Rhone Valley (France) and Australian GSM blends (Grenache/Syrah/Mourvèdre). American oak also imparts the spicy tones of cloves and vanilla to several red wines from California and Mexico, in particular those based on Zinfandel and Petite Sirah.

### CREATIVITY WITH EUGENOL

As part of an event called "A Molecular Tasting Meal with Two Master Chefs" (see the "Experiments in Food Harmony and Molecular Sommellerie" chapter), which took place at the restaurant L'Utopie in Quebec City in March 2009, I presented some results of my work on food and wine pairings.

On that occasion, the menu was inspired by aromatic molecules that I had selected in collaboration with the Bordeaux wine scientist Pascal Chatonnet. It was built around foods complementary to these molecules, which I had discovered during the course of my research.

The main course was conceived "for and by" a Château de Beaucastel 2005, which inspired an aromatic theme of complementary foods based on dimethylpyrazine (cocoa) and eugenol (cloves). The foods harmonized beautifully with this great red Châteauneuf-du-Pape, with its clearly defined notes of cocoa and cloves. The carefully selected food included **Inuit caribou, cooked in its juices with blackberry seeds, two types of celeriac purée (licorice and clove), honey fungus with cacao nibs, and Thai basil leaf.**

### CLOVES AT THE TABLE

In Europe, cloves are used mainly to enhance the taste of desserts, while in most other parts of the world they serve to render meat's flavor more complex. The main use of cloves, however, is to perfume the well-known Indonesian cigarettes called *kreteks*—95% of world clove production is for this purpose!—which contain about 40% cloves.

We find cloves in classic recipes such as gingerbread, sauerkraut, pineapple-glazed ham (cooked pineapple also contains eugenol; see the "Pineapple and Strawberries" chapter), the traditional Quebecois pig trotter stew (*ragoût de pattes*

## 2. COMPLEMENTARY FOODS
**CLOVES**

APRICOTS
BEER
CHINESE ROSEBUDS
CINNAMON
COFFEE
COOKED ASPARAGUS
FIVE SPICES
FOUR SPICES
GINGERBREAD

GRILLED BEEF
HAZELNUTS
INDIAN CURRY
MALTED BARLEY
MANGOS
MAPLE SYRUP
MOZZARELLA
PINEAPPLE
THAI BASIL

TOAST
TOASTED COCONUT
ROSEMARY
SCOTCH
STRAWBERRIES
VANILLA
WILD BASIL

de cochon), Indian curries, and flavorful boiled meats, such as pot-au-feu (French beef stew).

Cloves are also part of five-spice powder, which is most popular in Asian cuisine but well known around the world. This mixture most commonly includes badian (star anise), Sichuan pepper, cinnamon, fennel seeds, and cloves. Moreover, cloves almost always dominate the other spices.

Finally, cloves feature to varying degrees in other spice mixtures, such as ras el hanout (Morocco), garam masala (India), recado rojo (Mexico), and quatre-épices (France, along with pepper, nutmeg, and cinnamon).

Chinese rosebuds (sold dried) taste of pepper and cloves. So we may add these fragrant buds to dishes that feature cloves, or even replace the cloves in a recipe with them!

## PINEAPPLE, STRAWBERRIES, ROSEMARY, VANILLA: MOLECULAR COUSINS?

Pineapple, strawberries, rosemary, and vanilla all contain eugenol—especially noticeable in the case of very ripe or cooked pineapple and strawberries—which brings them closer to cloves and their accompanying wines.

So logic—and practice—dictate that a dish containing pineapple, strawberries, rosemary, vanilla, or cloves will be a harmonious match with wines marked by eugenol (see the "Pineapple and Strawberries" chapter).

## SPICE UP YOUR CHEESES!

Does the pairing of tannic red wines and cheeses almost always leave you, as it leaves me, with a disagreeable bitter taste in your mouth? That's normal! White wines do much better at this game (see the chapter on cheeses), even if you persist in looking at the cheese plate with "red"-colored glasses…

Given the understanding we have today of the major aromatic compounds comprising the fragrances of cloves and of wines, however, it is possible to prepare cheeses in a way that enhances their pairing with red wine.

### SOMMELIER-COOK'S HINT

**Bloomy-rind cheese perfumed with cloves** A simple and efficient trick to help your favorite red wine stay in the saddle… Choose a bloomy-rind cheese, such as Camembert or Brie. Cut the cheese horizontally into two pieces. Sprinkle very finely crushed cloves on top of the first piece of cheese, the piece on the bottom. Place the second piece back atop the first piece, wrap, and let macerate for several days in a cool location. You can replace the cloves with Chinese rosebuds, Thai basil, or Chinese five-spice powder and you'll obtain the same harmonious results!

### DUCK FAT ENFLEURÉ…

Once the molecular structure of a spice such as cloves is identified, ideas come together! Enfleurage, a practice formerly used by the perfume industry to capture the delicate fragrances of flowers in fatty substances, can be adapted for the kitchen as a new way to spice up food.

### SOMMELIER-COOK'S HINT

**Duck cooked in fat enfleuré with cloves** Take hot, but not boiling, duck fat and add several cloves (or a complementary spice). Let the mixture macerate in the refrigerator for several days so that the fragrances are captured in the fat (as was done in the old days for the enfleurage of rose petals). Then, after removing the cloves, lightly heat the duck fat.

After this process, you'll come out with duck fat with the subtle taste of cloves, which you can then use to spice up your foods! Duck confit and sautéed potatoes will never have seemed so inspiring, and the harmony of ingredients cooked in this duck fat will be even better when paired with wines possessing the same aromatic characteristics.

To create beautiful harmonies with other types of wines, try selecting different spices (ajowan, cinnamon, coriander seeds, nutmeg, nigella, Sichuan pepper, licorice, rosemary, saffron). Explore the possibilities of this simple and effective culinary technique, inspired by past practices of the perfume industry. It will make your recipes more fragrant and consequently allow for more subtle wine and food pairings.

BLOOMY-RIND CHEESE PERFUMED WITH CLOVES
AND DRIED ROSEBUDS

## THE SPICE ROUTE

As I mentioned in the "Sotolon" chapter, in 2007 I had the great privilege of inspiring a menu with the theme "The Spice Route" at chef Racha Bassoul's former Montreal restaurant, Anise.

The fourth act of this meal is a good example of how cloves can work in harmony with other ingredients and with wine:

**Villa de Corullón 2001 Bierzo,**
*Descendientes de J. Palacios, Spain*
**and Pot-au-Foie**
*Under a canopy of duck foie gras and cloves*
*infused with Chinese star anise,*
*curry leaves, and Sichuan pepper*
*and rice flavored with wulong Ali Shan tea, 1993*

Our goal was to create a vibrant aromatic harmony between the original foie gras stew and the Bierzo, which is based on the very rare Mencia grape. This native Spanish grape variety, resuscitated by the illustrious Álvaro Palacios and endowed with subtle fragrances of cloves, vanilla, and smoke, paired beautifully with the dish, as we had made good use of the fragrant, sensual clove when macerating the foie gras.

To round off this aromatic pairing, we replaced the original green tea that we had used in this dish at its creation the previous year with a smoked black tea, Ali Shan 1993, whose smoky and woody fragrances are reminiscent of cloves and of the wine we had chosen. The subtle peppery tones of the curry leaves and of the Sichuan pepper, also found echoes in the wine. And to finish the dish, what could be better than a roll perfumed with black tea?

## ÁLVARO PALACIOS, A WINEMAKER OF MANY TALENTS

The magical touch of Álvaro Palacios propelled him, at the end of the 1980s, to the forefront of the renaissance of the Priorat appellation zone, as director of the wine estate that carries his name. Since then, he has also helped spearhead another renaissance: that of the Mencia grape variety in Bierzo, where, along with his nephew Ricardo Pérez Palacios, he founded the wine estate Descendientes de J. Palacios. As if this were not enough, in 2000, after the death of his father, he returned to take the reins of Palacios Remondo in Rioja, where he was able to revitalize the vintages of his family wine estate. Supported by his wife Cristina, this young, gifted winemaker is definitely a Renaissance man!

## SOME DISHES THAT PAIR WELL WITH POWERFUL REDS

**Priorat, Rioja, Zinfandel, and Petite Sirah**

+ Lamb stew with four spices (pepper, nutmeg, powdered ginger, and cloves)
+ Lamb tajine with five spices and caramelized cippolini onions
+ Salmon steak with black coffee and five Chinese spices (recipe in *À table avec François Chartier*)
+ Lamb pot-au-feu (cooked rare) with tea and spices (star anise, licorice, cinnamon, cardamom seeds, cloves, and black tea leaves)
+ Cincinnati chili

## SOME DISHES THAT PAIR WELL WITH MODERATE TO VIGOROUS REDS

**Bierzo, Campo de Borja, Cariñena, New World Pinot Noir, and Rhone Valley or Languedoc Grenache**

+ Grilled salmon fillet covered with crushed Chinese four spices (pepper, nutmeg, powdered ginger, and cloves)
+ Risotto with clove-flavored beet juice
+ Quail glazed with honey and five spices (may be accompanied by risotto with clove-flavored beet juice)
+ Roast chicken with sesame seeds and five spices
+ Christmas turkey accompanied by risotto with clove-flavored beet juice
+ Beet salad flavored with four spices and olive oil (without vinegar)

## DESSERT WINES

A dessert made with strawberries or pineapple—fruits that share the exact same major compounds (see the "Pineapple

3.

# COMPLEMENTARY WINES

CLOVES

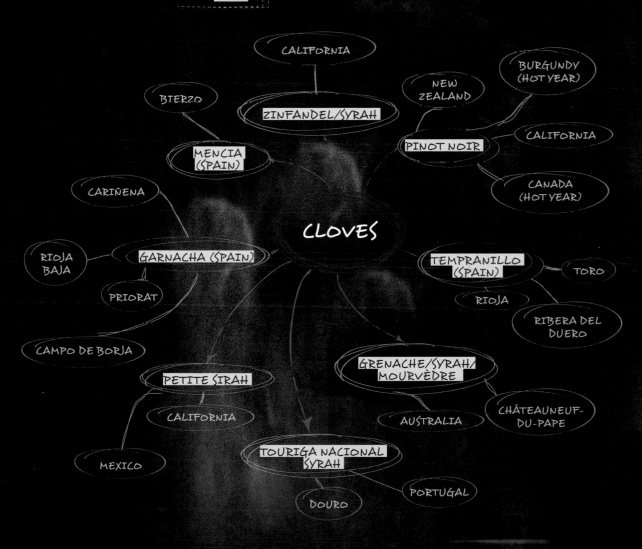

CALIFORNIA

BIERZO

ZINFANDEL/SYRAH

NEW ZEALAND

BURGUNDY (HOT YEAR)

PINOT NOIR

CALIFORNIA

CANADA (HOT YEAR)

MENCIA (SPAIN)

CARIÑENA

CLOVES

RIOJA BAJA

GARNACHA (SPAIN)

TEMPRANILLO (SPAIN)

TORO

PRIORAT

RIOJA

RIBERA DEL DUERO

CAMPO DE BORJA

PETITE SIRAH

GRENACHE/SYRAH/ MOURVÈDRE

CALIFORNIA

AUSTRALIA

CHÂTEAUNEUF- DU-PAPE

MEXICO

TOURIGA NACIONAL SYRAH

PORTUGAL

DOURO

and Strawberries" chapter)—as well as rosemary, vanilla, or cloves, finds its Holy Grail of pairing harmony in dessert wines endowed with fragrances from the eugenol family, such as:

+ Banyuls, Maury, Rasteau, and Rivesaltes (young)
+ Monastrell Dulce, Jumilla, Spain (young)
+ Pedro Ximénez "Solera" Montilla-Moriles, Spain
+ Red Pineau des Charentes, five to ten years old
+ Vintage port (young)

## SOME DESSERT RECIPE IDEAS

+ Strawberries with pepper and cloves
+ Gingerbread and pineapple millefeuille (a puff pastry with a layered filling). You can find the recipe in *À table avec François Chartier;* modify the syrup to contain more cloves and a touch of vanilla.
+ Gingerbread and strawberry millefeuille
+ Pineapple shortcake with rosemary-flavored whipped cream
+ Strawberry shortcake with rosemary-flavored whipped cream
+ Warm pineapple and strawberry soup with rosemary
+ Pineapple tarte tatin with cloves and vanilla ice cream

JUNIPER

FIR

PINE

PINENE

SAGE
SPRUCE BEER
STRAWBERRIES
TURMERIC
VERBENA

# ROSEMARY

## A SOUTHERNER... WITH AN ALSATIAN PROFILE!

"The introduction of new elements to the domain of knowledge must first of all be considered temporary and should be fully criticized and justified."

HUBERT REEVES

The above quotation from the famed Quebecois astrophysicist Hubert Reeves confirms once again that in science, as in the arts of the table, experimentation leads to the kind of thinking that nurtures creativity—which, in turn, can open the door to many pleasures. In this chapter, we'll follow the sunny and woody aromatic path of rosemary, and of those wines that harmonize perfectly with this great Mediterranean herb.

### CHEMISTRY 101

Let's start with some basic chemistry. Rosemary's volatile compounds belong to the terpene family. Terpenes are substances of primarily plant origin that typify the fragrances of some wines, in particular those based on Muscat and, to a lesser extent, Gewürztraminer and Riesling; these are all endowed with terpenic fragrant molecules. So we're talking about white wines. However, in most cases, rosemary is used in cooking to enhance meat dishes, such as lamb, which the great majority of amateurs and professionals inevitably pair with red wines.

And when rosemary is used in a vegetable or fish dish, the principle of choosing wine according to the food's geographic origin almost always leads to a white wine, or to a rosé from Provence or from a vineyard in the Mediterranean basin.

On the other hand, Riesling- and Gewürztraminer-based white wines are nearly absent from the vineyards of southern France. As for Muscat, it is native to this region and is

very abundant, but it is almost always vinified into a naturally sweet wine (VDN, *vin doux naturel*), syrupy and rich in alcohol. So Muscat wines are not really suitable for pairing with such savory dishes, with the obvious exception of some fine Alsatian dry Muscat wines.

You may be thinking that you have rarely, if ever, sensed a hint of fresh rosemary in a Riesling or Gewürztraminer. But bear in mind that the fragrances of herbs and spices, just like those of vegetables and animal products, are characterized by more than one aromatic compound.

In certain cases, one compound dominates the others, such as cloves, cinnamon, star anise, or thyme. But 99% of the time, it is the ensemble of compounds that gives each herb or spice its particular aroma. Coriander seeds, for example, are both floral and lemony due to their many molecules from the floral and citrus families. Rosemary is no exception; it proves to be quite complicated, aromatically speaking.

The next time that you have a sprig of rosemary in your hands, breathe in its fragrance deeply. You will discover both floral and woody notes, but also touches reminiscent of conifers, cloves, and eucalyptus.

These fragrances, which constitute rosemary's unique and widely recognizable bouquet, come from various volatile compounds belonging to the terpene family—chemical molecules that are naturally produced by plants in self defense, to repel animal predators.

This brings us closer to the fragrances associated with Riesling and Gewürztraminer. As proof, consider that it is common to smell the fragrances of flowers and conifers in Riesling, just as it is common to discover in a Gewürztraminer wine spicy notes of cloves, as well as camphor-smelling notes of eucalyptus and floral rose notes. That explains the symbiotic union between rosemary and these wines—not only in theory, but also confirmed in practice by our senses.

Finally, there are aromatic differences among different rosemary varieties: Corsican rosemary is richer in borncol (which has a camphor, woody, tonic, and almost medicinal fragrance), with soft, fruity fragrances and a touch of incense. Continental rosemary, from Provence, the southern Rhone Valley, and Languedoc-Roussillon, is characterized by verbenone, which has a fragrance of Spanish verbena.

## TERPENES

Terpenes, which make up the majority of rosemary's volatile compounds, are complex aromatic compounds that tend to have floral nuances. This class of hydrocarbon, which is produced by many plants, such as conifers, is a major constituent of the floral and citrus aromas that dominate Muscat-based wines and Gewürztraminer, which also contains the family of hydrocarbons found in Riesling wine.

More than four thousand terpenic compounds have been compiled; this includes some four hundred monoterpenes and approximately one thousand sesquiterpenes. These hydrocarbons are extracted from essential oils and vegetal resins. The most important terpenes are alpha-pinene, beta-pinene, delta-3-carene, limonene, carotene, and lutein.

### TERPENIC GRAPE VARIETIES

Grape varieties of the Muscat family are the most richly endowed with terpenes, even if other grape varieties and other wines are also dominated by the different volatile compounds of the terpene family.

Terpenes occur only in white grapes, with the exception of Black Muscat. They possess the aromatic characteristics also found in conifer needles and bark, as well as in citrus fruit. Terpenes express the fresh tonalities of spruce, citrus fruit, flowers, and green leaves.

In 1956, the scientist R. Cordonnier discovered terpenes, as well as their role in typifying Muscat's aroma. Since then, a great number of researchers and wine scientists have studied these volatile compounds in order to understand better their impact on wines from the large family of Muscats, as well as on wines from other white grape varieties.

The volatile substances of all cultivated wine grape varieties belong principally to two large families of volatile, aromatic compounds: terpenes and pyrazines.

As we have seen, terpenes are responsible for the flower and citrus fruit fragrances that dominate the Muscat family, as well as Gewürztraminer and its fraternal twin, Scheurebe (see the "Gewürztraminer..." chapter). We can also add to this the family of hydrocarbons whose spruce, pine, rosemary, fir, and gasoline fragrances are found in Riesling wine.

Other grape varieties also develop terpenic notes, to greater or lesser degrees; sometimes they are quite subtle. These varieties include Albariño and Viura from Spain; Müller-Thurgau from Austria; and the very French Chardonnay, Muscadelle, Roussanne, and Sauvignon Blanc.

These terpenic tonalities are conveyed by volatile compounds such as linalool (floral/fruity), geraniol (rose/woody/spicy), nerol (floral/citrus), hotrienol (linden/lavender, ginger/fennel, and honey), and alpha-terpineol (citrus).

## THE MOST COMMONLY OCCURRING TERPENIC FRAGRANCES

Bergamot, camphor, cinnamon, citrus, eucalyptus, gasoline, ginger, hibiscus, lavender, lemongrass, lily of the valley, mint, nutmeg, pine, rosemary, rosewater, rosewood, saffron, sage, spruce, sweet peach, thyme, woody and spicy notes, and ylang-ylang.

Grapes and wines contain more than seventy terpenic compounds. For the most part, they are monoterpenes, but there are also some sesquiterpenes and corresponding alcohols and aldehydes.

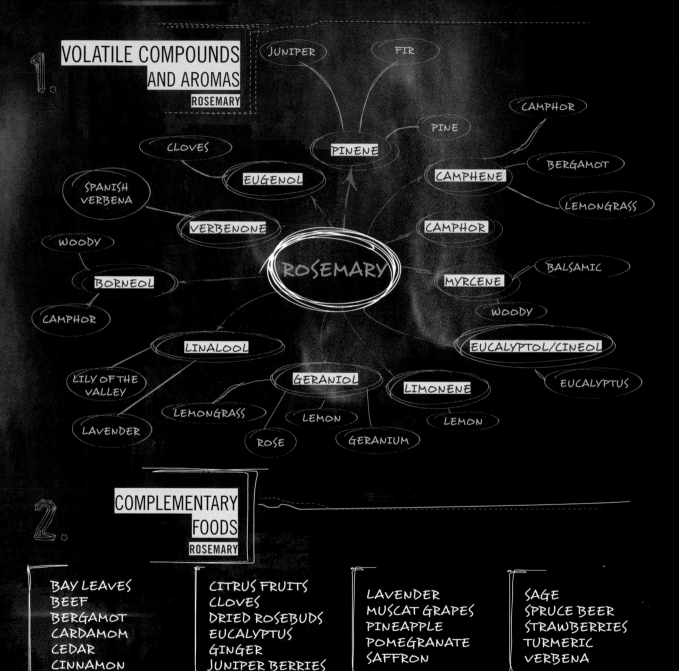

# 1. VOLATILE COMPOUNDS AND AROMAS
ROSEMARY

ROSEMARY

- PINENE
  - JUNIPER
  - FIR
  - PINE
- CAMPHENE
  - CAMPHOR
  - BERGAMOT
  - LEMONGRASS
- CAMPHOR
- MYRCENE
  - BALSAMIC
  - WOODY
- EUCALYPTOL/CINEOL
  - EUCALYPTUS
- LIMONENE
  - LEMON
- EUGENOL
  - CLOVES
- VERBENONE
  - SPANISH VERBENA
- BORNEOL
  - WOODY
  - CAMPHOR
- LINALOOL
  - LILY OF THE VALLEY
  - LAVENDER
- GERANIOL
  - LEMONGRASS
  - ROSE
  - LEMON
  - GERANIUM

# 2. COMPLEMENTARY FOODS
ROSEMARY

- BAY LEAVES
- BEEF
- BERGAMOT
- CARDAMOM
- CEDAR
- CINNAMON

- CITRUS FRUITS
- CLOVES
- DRIED ROSEBUDS
- EUCALYPTUS
- GINGER
- JUNIPER BERRIES

- LAVENDER
- MUSCAT GRAPES
- PINEAPPLE
- POMEGRANATE
- SAFFRON

- SAGE
- SPRUCE BEER
- STRAWBERRIES
- TURMERIC
- VERBENA

### MUSCAT AND *BOTRYTIS CINEREA*

When Muscat grapes are attacked by *botrytis cinerea* (noble rot) at a certain level of intensity, they lose their typicity, due to a significant reduction in terpenes. This means that the grapes lose their unique floral character, which explains why most Muscat wine growers don't want to see *botyrytis cinerea* develop in their vineyards.

Terpenic aromatic molecules are very volatile. They are therefore the first molecules we perceive in the nose and in the mouth when we savor a wine—think of the very immediate smell of gasoline in some Rieslings, or that of roses in Gewürztraminers. Moreover, their volatility makes them disappear fairly quickly during cooking, both in the wine and in aromatic herbs (such as rosemary), citrus fruit, and flowers.

### ROSEMARY AT THE TABLE

**SHRIMP WITH ROSEMARY**

In a frying pan, sauté some diced pineapple and red pepper, as well as some finely chopped fresh rosemary sprigs. Deglaze with a white wine, the one you will be serving with the meal.

Add shrimp and brown it, and add a bit of cream to the mixture. If you have deglazed with a dry Riesling, serve the same Riesling. If you have chosen a dry Gewürztraminer instead, you can make the food and wine pairing even more vibrant by accompanying the dish with a teaspoon of whipped cream enhanced with a pinch of turmeric, which also pairs beautifully with Gewürztraminer.

### A STEAM ROOM WITH ROSEMARY VAPORS...

To create an even stronger aromatic link between the fragrances of rosemary and those of your selected wine, serve this dish in a bowl placed in a larger deep dish containing several whole sprigs of fresh rosemary. When it's time to serve your guests, pour a bit of boiling water into the deep dish (for more details, see the *"Fricassée de crevettes..."* recipe in *À table avec François Chartier*). The rosemary scent will waft into your guests' noses, as if they were in a steam room with rosemary vapors. Your guests need only let themselves be guided by the rosemary's aromatic compounds towards the wine in their glasses. You'll thus create the desired union—both "scientific" and gourmet—between rosemary and Riesling or Gewürztraminer white wines.

## 3. COMPLEMENTARY WINES
**ROSEMARY**

ALBARIÑO (SPAIN)
BLACK MUSCAT
CABERNET SAUVIGNON (AUSTRALIA/CALIFORNIA/CHILE)
FINO SHERRY
GEWÜRZTRAMINER
GRENACHE NOIR (SPAIN/LANGUEDOC/RHONE VALLEY)

MÜLLER-THURGAU (AUSTRIA/ITALY)
MUSCAT
RIESLING
SCHEUREBE (AUSTRIA)
VIURA (SPAIN)

## ROSEMARY, LAMB, AND... RIESLING!

Now, what to do with meat and rosemary? Be daring and try boiling lamb in a stew, spiced up with several sprigs of rosemary, and surprise your guests by pairing this red meat with a noble dry Alsatian white Riesling wine. You'll attain harmony thanks to the two poles of harmonic attraction (see the "From Raw to Cooked" section of the "Beef" chapter).

First, boiled meat loses its blood and becomes stringier and very fragrant from the bouillon's flavors, opening the door to pairing with a dry, very fragrant white wine (or with a fruity, biting, very light red wine). Second, the presence of rosemary's major active compounds creates an almost perfect link with those compounds found in dry Alsatian Rieslings. In many cases, German Rieslings have an insufficient alcohol content, and so lack body and density, or may be a bit sweet.

## ROSEMARY: FROM THE CHEESE PLATE TO DESSERT

When it's time to serve the cheese plate, consider reuniting these same aromatic molecules. Serve a washed-rind cheese, such as Munster, in whose center finely chopped rosemary has macerated for several days. Accompany it with a late harvest Alsatian Gewürztraminer. This is a modern twist on the regional pairing of Gewürztraminer and the Alsatian specialty *Munster au cumin*.

And for dessert, the same type of wine will be a sensation with a rosemary-flavored pineapple and strawberry soup, or a surprising strawberry and pineapple shortcake topped with rosemary-flavored whipped cream.

### SOMMELIER-COOK'S HINT

**Siphoned lemon meringue pie** Why not try using a siphon to put a contemporary twist on the classic lemon meringue pie? In this case, the meringue is prepared with a whipped cream siphon, which makes it more airy, and perfumed with rosemary.

**Two desserts uniquely inspired by terpenic compounds:**

+ Campari and rosewater jelly prisms with pineapple, strawberries, clove-scented citrus juice, tofu steamed with rosemary vapor, eucalyptus sherbet, and ginger sticks (idea by François Chartier; see photo)
+ Pomegranate with Muscatel, citrus confit, and eucalyptus

ice cream (see recipe in *Spain Gourmetour*, No. 62, page 124). When preparing the plate, add rosemary and flowers such as jasmine and lemon verbena.

These desserts are 100% terpenic! So they are made-to-order for a late-harvest Riesling, a Gewürztraminer, or a sweet Muscat.

As mentioned previously (see the "Fino and Oloroso" chapter), fino sherry is also a good companion for rosemary. It is rich in terpenic floral notes (linalool, nerolidol, and farnesol), which give it an aromatic power and a presence in the mouth that make it a good match for the vivacity of rosemary. So a fino sherry pairs very well with, for example, a goat cheese salad marinated in rosemary olive oil.

## PINEAPPLE, STRAWBERRIES, AND CLOVES: IN ROSEMARY'S AROMATIC SPHERE OF INFLUENCE

As we discovered in the "Pineapple and Strawberries" and "Cloves" chapters, pineapple and strawberries (especially when very ripe), cloves, and rosemary all contain a good dose of eugenol. So a dish containing any or all of these ingredients will be a harmonious match for wines characterized by eugenol.

## SAGE, TURMERIC, BAY LEAVES, EUCALYPTUS...

Just like rosemary, sage and turmeric are rich in terpenic notes, and thus both pair well with Riesling, Muscat, and Gewürztraminer wines.

SIPHONED LEMON MERINGUE PIE

The essential oils of eucalyptus and of bay leaf show only minute differences in their active components. We can thus qualify them as molecular twins, similar to pineapple and strawberries. Cineol (or eucalyptol) and borneol, the two major active components, provide the tone for each of them. These two volatile compounds are also found in rosemary. This explains the successful pairing of dishes flavored with eucalyptus, bay leaves, or rosemary and red wines with a eucalyptus nose, such as some Chilean, Californian, and Australian Cabernet Sauvignons.

## NEW PATHWAYS TO CREATIVITY IN THE KITCHEN

+ Shrimp fricassee with pineapple and sweet peppers, topped with rosemary whipped cream (recipe in *À table avec François Chartier*) **German, Alsatian, or Australian Riesling**

+ Shrimp fricassee with pineapple and sweet peppers, topped with curried whipped cream and flavored with rosemary **Dry Alsatian Gewürztraminer**

+ Lamb pot-au-feu flavored with rosemary **Dry Alsace Grand Cru Riesling**

+ Washed-rind cheese flavored with rosemary and warm pineapple and strawberry soup flavored with rosemary **Gewürztraminer Vendanges Tardives (late harvest)**

+ Strawberry and pineapple shortcake with rosemary whipped cream **Muscat de Rivesaltes (or another naturally sweet wine based on Muscat)**

+ "Siphoned" pie (lemon pie with siphoned meringue flavored with rosemary) **Riesling Vendanges Tardives (late harvest)**

BITTER TASTE

PICROCROCIN

CINNAMON
(CEYLON/SRI LANKA/
CINNAMOMUM ZEYLANICUM)

RIESLING
BLACK AND
GREEN TEA

# SAFFRON

## THE QUEEN OF SPICES

> "Only an audacious reflection and not an accumulation of facts can help us to progress."
> ALBERT EINSTEIN

We move now to the hot and intriguing volatile compounds of saffron, its complementary ingredients, and the wines that work in harmony with this queen of spices from the Mediterranean basin.

Saffron is the most expensive naturally occurring spice. It takes some two hundred thousand mauve crocus flowers—each of whose three long blood-red stigmata are removed by hand—to produce a single kilogram of saffron. More than forty hours of work—harvesting, separating the stigmata, and drying (which involves roasting them on the embers of a fire without smoke or flames)—go into producing a single kilogram. Now you know where it gets its reputation as the Queen of Spices!

Saffron was originally domesticated in Greece, during the Bronze Age. Later, it was transported by caravans as far east as Kashmir. Then the Arabs distributed it in the West, as far as Spain. Finally, during the Crusades in the Middle Ages, our queen spread into France and England.

Today, Kashmir and Iran are the two most important saffron producers. However, it is Spanish companies, almost all of them located in Novelda, Alicante—headquarters of Verdú Cantó Saffron Spain, the international point of reference for the industry—that are responsible for 90% of the world trade of this delightful spice.

### SAFFRON IN CLASSIC DISHES AROUND THE WORLD

For centuries, saffron has been used in Iran and in Spain to flavor rice dishes such as pilaf and paella. The well-known Italian dish risotto alla Milanese carries the aromatic signature of saffron, as does the fish stew from Marseille, France, called bouillabaisse. The Indians, meanwhile, use the spice in sweet dishes, as well as in the salty-spicy world of certain curries.

### MORE THAN 150 VOLATILE COMPOUNDS

Saffron gives off a hot, penetrating floral and spicy fragrance, vaguely reminiscent of dry hay, while developing a bitterness in the mouth that can vary in intensity, as well as a lightly piquant note.

Of course, there are several varieties: oriental saffrons, which are dried in the sun on large trays, are rather spicy and less "saffrony," while European saffrons, which are oven-dried, seem more "saffrony," with classic touches of flowers and dry hay.

Saffron from Sussex, for example, has a sweet, seductive fragrance that comes from a combination of aromas, including oranges, blond tobacco, and Indian tea—with a touch of wet cardboard!

Saffron's yellow-orange color comes from molecules originating in the decomposition of carotenoids, such as alpha-crocin, a diester that also contributes to saffron's singular bouquet, as you will read later in this chapter.

Although it is commonly believed that the fragrance of an herb or spice is singular—in other words, composed of a single aromatic molecule that gives the spice its specific character—this is not the case. Quite on the contrary, the aroma of each herb and each spice contains a cocktail of volatile compounds that, when mixed together, create its characteristic final aroma. But generally, only a handful of these aromatic molecules dominate the others.

In the case of saffron, more than 150 volatile molecules contribute to its unique fragrance, of which about thirty constituents are fairly important. Of these, fewer than a dozen compounds dominate, including pinene (with the smell of pine and fir trees), cineol (also called eucalyptol, the major compound in eucalyptus and cardamom), and especially safranal.

Numerous non-volatile compounds also contribute to saffron's aromatic structure and strongly color the liquid when saffron is immersed in hot water, milk, cream, or alcohol.

## INFUSE BEFORE COOKING

It is essential to hydrate saffron in a hot or warm liquid prior to use to bring out its color and fragrance, whose intensity is directly proportional to the infusion time. A period of thirty minutes is recommended (be careful, though: excessive infusion can produce a very bitter, invasive taste). A fatty liquid (for example, butter, milk, cream, fatty bouillon, or oil) or a liquid containing alcohol are better choices than water for maximum extraction; saffron's precious aromatic compounds are soluble only in fatty liquids or in alcohol, and therefore very little (if at all) in water.

## SAFRANAL AND PICROCROCIN

Among the compounds that contribute the most to saffron's signature taste is picrocrocin, a bitter-tasting carotenoid that is formed from the union of a glucide and an aldehyde.

## SAFFRON'S CAROTENOIDS IN OTHER INGREDIENTS

The same carotenoids found in saffron are also found in yellow apples, quinces, grapes, tobacco, roses (as well as dried rosebuds and rosewater), Australian boronia flowers, and osmanthus, a Chinese flower used to perfume tea (see the "Experiments in Food Harmony and Molecular Sommellerie"

chapter). So these ingredients work in harmony with saffron on the plate, and they also pair well with wines in the same aromatic sphere as saffron.

Safranal, or 2,6,6-trimethyl-1,3-cyclohexadiene-1-carboxaldehyde, a volatile terpene that is less bitter than picrocrocin, gives saffron its distinctive fragrance. It is also a powerful antioxidant. Safranal represents 70% of saffron's total volatile compounds. It also occurs in black tea, mate, paprika, pimentón (Spanish paprika), pink grapefruit, and osmanthus flowers. All of these foods are thus a good choice to cook alongside saffron, and they can all be successfully paired with wines suggested for this Queen of Spices.

A third compound, 2-hydroxy-4,4,6-trimethyl-2,5-cyclohexadien-1-one, also plays a part in saffron's aroma, bringing to it notes of dry hay. It is the compound that contributes the most to the fragrance of some saffron varieties, such as red saffron from Greece, even though safranal provides up to 70% of the volatile molecular composition of most types of saffron.

## PINENE: A REACTIVE MOLECULE

Pinene, which has a fragrance of pine, fir, and juniper, is also among the dominant volatile compounds in saffron's aromatic structure. Interestingly, it is soluble in alcohol but insoluble in water.

To extract pinene's fragrance from saffron, as from any other spice characterized by this molecule, the saffron must enter into contact with alcohol or another solvent that can extract it.

Pinene also occurs in many other herbs and spices, in two different forms: alpha-pinene (in ginger, lavender, mint, sage, and thyme) and beta-pinene (in yarrow, basil, parsley, rosemary, and roses). These two compounds also occur in the aromatic profile of juniper berries, spruce beer, cinnamon, and, of course, saffron.

These ingredients open up new paths to making saffron-based dishes more complex and harmonious, and to creating successful pairings with wines in the same aromatic sphere.

Note that the greater the number of ingredients containing pinene there are in a given dish, the more this tonality will dominate when it comes to savoring it with a wine.

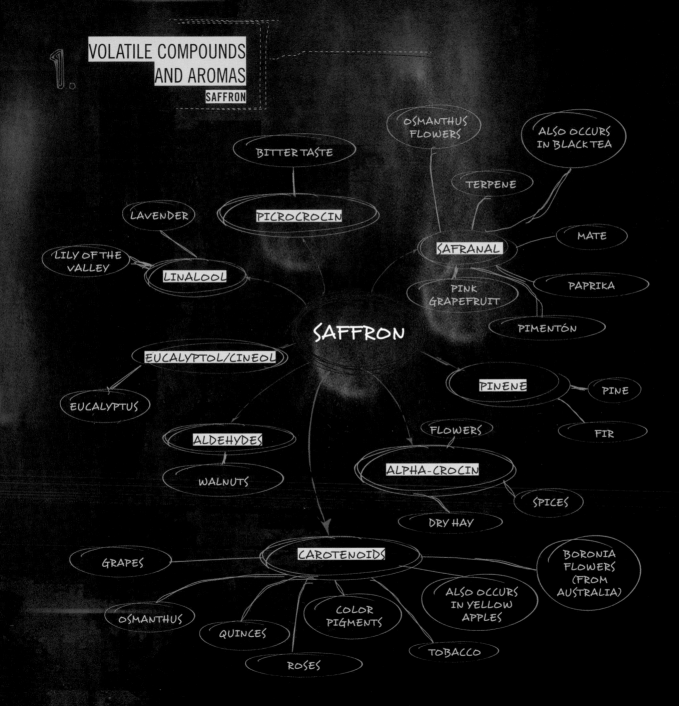

OSMANTHUS FLOWERS

ALSO OCCURS IN BLACK TEA

BITTER TASTE

TERPENE

LAVENDER

PICROCROCIN

MATE

SAFRANAL

LILY OF THE VALLEY

LINALOOL

PINK GRAPEFRUIT

PAPRIKA

SAFFRON

PIMENTÓN

EUCALYPTOL/CINEOL

PINENE

PINE

EUCALYPTUS

FLOWERS

FIR

ALDEHYDES

ALPHA-CROCIN

WALNUTS

SPICES

DRY HAY

GRAPES

CAROTENOIDS

BORONIA FLOWERS (FROM AUSTRALIA)

ALSO OCCURS IN YELLOW APPLES

OSMANTHUS

COLOR PIGMENTS

QUINCES

TOBACCO

ROSES

Another interesting fact: pinene, in particular alpha-pinene, is very reactive to iodine, and thus sublimates the taste of the latter.

## PINENE AND IODINE

If you cook iodine-tasting seafood with saffron or other ingredients rich in alpha-pinene, the iodine note will dominate the dish's flavor. This is because alpha-pinene has the capacity to sublimate iodine's flavor, making it more powerful. Likewise, in wine pairings, the iodine taste of such seafood will intensify if the wine served has powerful pinene aromas. This explains the beautiful union between Riesling and oysters or scampi.

## SAFFRON AND WINES

To choose a wine that will successfully pair with dishes dominated by saffron, let's begin by noting that carotenoids, aldehydes, and terpenes are the three principal families of volatile compounds that contribute to saffron's fragrance and taste.

Some of these compounds give it a bitterness that must be taken into account when selecting a wine to accompany a saffron-flavored dish. Logically, one must go with wines rich in these three types of volatile compounds.

## RIESLING, SAUVIGNON BLANC, CHARDONNAY...

Carotenoids are found especially in "non-aromatic" grape varieties, including Chardonnay, Sauvignon Blanc, and Riesling. But because terpenes are also, to a large extent, among the aromatic molecules of Riesling, in particular monoterpenes—which, like saffron, present spicy and floral tones—the choice of Riesling is almost automatic!

Linalool, Riesling's principal terpene, which has a floral aroma, finds a powerful echo in saffron's terpenic florality. Linalool is also the major aromatic compound in lavender, bergamot, mint, citrus fruits, Ceylon (Sri Lanka) cinnamon, sweet European basil, and fresh figs.

Note that the best Riesling wines usually present a strong minerality at the finish in the mouth, generating a soft impression of bitterness, as does saffron.

Thus, a successful pairing with a saffron-spiced monkfish kebab, or a dish of monkfish flavored with onions, lemon, and saffron, requires a white wine that's simultaneously dense and very vivacious, with aromas from the same sphere of aromatic compounds as saffron.

One good selection is the Léon Beyer Riesling Les Écaillers 2003, Alsace, France. In this case, both weakness

IODINE TASTE

and overripeness are totally absent. There is only fruit, with touches of pink grapefruit and terpenic notes that bring to mind rosemary, saffron, spruce, and citrus fruits. This is a dry wine, straight and gripping, without being tense, but with lots of freshness. It has a lovely texture and surprising density in the mouth, and offers very long mineral flavors that brilliantly hold up to the dense, flavorful monkfish as well as the heady aromas of saffron.

## FINO, MANZANILLA, ROSÉ WINES...

Of course, some Chardonnay and Sauvignon Blanc wines also offer electrifying minerality in the mouth. Such is also the case for some dry Muscats and fino and manzanilla sherries, which, given their sweet bitterness and their terpenic compounds, as well as their richness in aldehydes (in the case of the sherries), successfully pair with some saffron-rich foods.

For example, try the delectably fragrant and subtly bitter Pierre Sparr Muscat Réserve, Alsace, France, as an aperitif to accompany canapés of shrimp with mayonnaise and saffron or creamed carrots with saffron and mussels.

Another harmonious pairing along these lines is a fino sherry, which, like saffron, is rich in aldehydes (aromatic compounds that appear during the wine's maturation), and a dish of creamed scallops and scampi flavored with saffron.

Such a pairing calls for the world-renowned Tio Pepe Fino Sherry, Gonzalez Byass, Spain, which has a great charm and an inviting vivacity, exhaling direct and simple notes of green apples and fresh almonds. In the mouth, it is crunchy, invigorating, thirst-quenching, and as expressive as can be.

An interesting fact to know, especially during the spring: many rosé wines are also rich in carotenoids, as is saffron. This explains, more scientifically, the lovely harmony that I have always found between rosés and dishes containing saffron!

Tempt your palate with cinnamon- and saffron-flavored lamb chops paired with a well-structured rosé, served cool rather than cold, such as Le Rosé de Malartic, Bordeaux Rosé, France. This vividly colored wine presents a very fine nose and inviting fruity notes. It is ample, textured, fresh, and almost greedy in the mouth, with a surprising floral presence (saffron). This is a serious wine!

## A CUP OF TEA?

It is also possible to create beautiful harmonies at the table by serving... a cup of tea! More precisely, Gyokuro green tea.

The unusual production process of Gyokuro green tea—three weeks prior to harvest, the tea plants are deprived of sunlight by bamboo arbors—stimulates the important development of theine (caffeine) and carotenoids.

In this way, Gyokuro green tea acquires a great complexity of volatile compounds from the carotenoid family, such as are found in saffron and the wines that pair well with it.

Gyokuro green tea should thus be served with dishes whose dominant taste is saffron, such as a risotto with peas and saffron; peas are rich in methoxypyrazines, plant molecules also found in green tea. So the pairing will be doubly molecular!

Don't hesitate to cook recipes in which Gyokuro green tea and saffron dominate; this will reinforce the harmonious link with wines in the aromatic sphere of saffron.

## THE YELLOW APPLE: A COUSIN TO SAFFRON

Like saffron, yellow apples such as Golden Delicious owe their yellow color to their richness in carotenoids, in particular beta-carotene. (Red apples owe their color mostly to anthocyanins, as does red wine.)

For a lovely wine and food pairing, create dishes containing yellow apples and saffron and serve them with the same wines that are suggested for saffron, which are also rich in carotenoids.

Good choices in this case would be wines based on Sauvignon Blanc and Chardonnay, which are in the same aromatic sphere as yellow apples and saffron. For example, cook a pork chop with Golden Delicious apples and saffron and serve a dry white Chardonnay-based wine, ideally one from a warm vintage year that was barrel-raised but is not over-oaked. Some Australian and Californian Chardonnays come to mind.

For a yellow apple cake with saffron, select a sweet white wine such as a late harvest Chilean Sauvignon Blanc, or a Sauternes whose blend contains more Sauvignon Blanc than Sémillon. Sémillon, when affected by botrytis cinerea (noble rot), as in the case of Sauternes, develops fragrances that are also similar to those of saffron. Such is also the case for

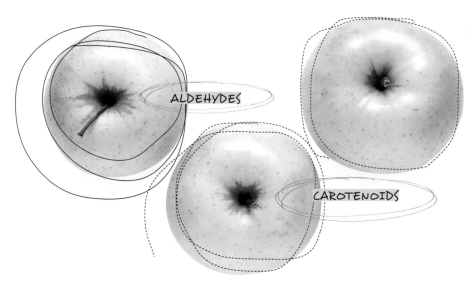

white wines from Jurançon based on Petit Manseng or Gros Manseng or both, whether dry or sweet.

## CATALAN FOOD AND WINE HARMONIES AT ELBULLI

**Apricots with vanilla and saffron, in an emulsion of green pistachios** (from elBulli's 2008 menu) A dish that goes very well with a fine Sauternes.

**Polenta gnocchi with coffee and saffron** (from elBulli's 2008 menu) In this case, go for an oaked Chardonnay, which simultaneously expresses the barrel's charred tonalities, also present in the coffee, and aromatic tonalities from the carotenoid family, also present in the saffron.

## BACK TO THE FUTURE

Traditionally, dishes containing saffron have been paired with either rosé wines or with Condrieu, a white wine from the Rhone Valley based on Viognier grapes. Condrieu was chosen both for its very aromatic character, as is the case for saffron, and for its generosity, which enables it to envelop the bitterness of saffron and last as long in the mouth.

But by understanding saffron's aromatic structure, as well as that of certain wines, it is now possible to refine our pairing approach and open up new avenues for both beginning cooks and professional chefs, for both wine amateurs and professional sommeliers.

0138. SAFFRON

## 2. COMPLEMENTARY FOODS
SAFFRON

BERGAMOT
BLACK TEA
BORONIA FLOWERS
CARDAMOM
CINNAMON

CITRUS FRUIT
DRIED LAVENDER
DRIED ROSEBUDS
EUCALYPTUS
FIR TREE BUDS
FRESH FIGS
GINGER

GOLDEN DELICIOUS
APPLES
GRAPES
GYOKURO GREEN TEA
JUNIPER BERRIES
LEEKS
MATE
MINT
OSMANTHUS FLOWERS
PAPRIKA
PARSLEY
PIMENTÓN
PINK GRAPEFRUIT

QUINCES
ROSEMARY
ROSEWATER
SAGE
SPRUCE BEER
SWEET EUROPEAN
BASIL
TOBACCO
YARROW
YELLOW APPLES

## 3. COMPLEMENTARY WINES AND TEAS
SAFFRON

CHARDONNAY
FINO AND MANZANILLA
SHERRIES
GYOKURO GREEN TEA

MATURE WHITE WINE
(CONTROLLED OXIDATION)

MUSCAT

RIESLING
ROSÉ WINE
SAUVIGNON BLANC

SCHEUREBE
AUSTRIA
NERAL
LIMONENE
FRUITY
BETA-BISABOLENE
SPICY
GERANIOL

# GINGER

## A SEDUCER WITH A GREAT POWER OF ATTRACTION

*"Non cogitat qui non experitur.* / He who does not experiment does not think."

MARGUERITE YOURCENAR

Ginger (*Zingiber officinale Roscoe*) is part of the Zingiberaceae family, which consists of more than seven hundred species. Its unique hot, musky, zesty, sugary, and strongly aromatic taste, simultaneously pungent, biting, refreshing, thirst-quenching, and spicy, can be reminiscent of the equally complex flavors of Gewürztraminer-based wines.

Zingiberaceae are rhizomes (roots) divided into about fifty genera. Spices in this family include galangal, grains of paradise (also called Guinea pepper or Melegueta pepper), cardamom, and turmeric.

Every year, more than 1.5 billion kilograms (about 3.3 billion pounds) of ginger are consumed across the globe, especially in Southeast Asia and India. In the west, it has recently gained in popularity due to the explosion of sushi bars and restaurants, where it is one of the major condiments along with wasabi and daikon. And let's not forget its age-old use in ginger cookies, gingerbread cake, ginger beer, and ginger ale.

### GINGER'S BROTHER AND CINNAMON'S COUSIN

Galangal, a rhizome from the same family as ginger, is stamped with ethyl cinnamate, an ester of cinnamic acid occurring in cinnamon, with a cinnamon/balsamic/honey fragrance and the sweet taste of apricots/peaches. Ethyl and methyl cinnamates are naturally present in strawberries and pineapple, Sichuan pepper, some varieties of basil, and the Australian *Eucalyptus olida* (strawberry gum). The latter contains a very high level of methyl cinnamate, and is used by the food and perfume industries to reproduce the aromas of strawberries and cinnamon. We can thus use galangal instead of ginger with cinnamon, strawberries, pineapple, Sichuan pepper, basil, and *Eucalyptus olida*.

### GINGER'S MOLECULAR STRUCTURE

Ginger contains several volatile compounds that contribute to its floral, citrus, woody, spicy, camphor, and cold-tasting tonalities, in the aromatic image of wines based on Gewürztraminer and its fraternal twin, Scheurebe (see the "Gewürztraminer/Ginger/Lychee/Scheurebe" chapter).

+ **Beta-bisabolene:** A sesquiterpene with a mostly balsamic fragrance, with notes of citrus and spices. This compound is also found in star anise, avocados, basil, cardamom, bergamot, limes, pine needles, and sandalwood.
+ **Beta-sesquiphellandrene:** One of the main volatile compounds of ginger, along with citral and nerolidol. This is a sesquiterpene whose fragrance is mostly woody but also herbaceous and fruity. It also occurs in galangal, turmeric, boronia flowers, and parsley.
+ **Camphene:** A hydrocarbon, more precisely a bicyclic monoterpene, that is cold-tasting. Camphene is a constituent of many essential oils, such as camphor, bergamot, and citronella.

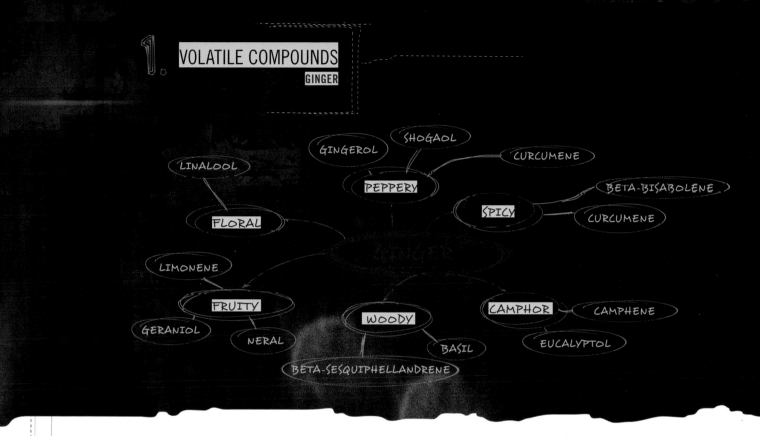

# 1. VOLATILE COMPOUNDS
## GINGER

+ **Citral (lemonal):** One of ginger's main volatile compounds, along with nerolidol and beta-sesquiphellandrene. It is also the major component of lemongrass, and is present in verbena, oranges, and lemons, among other foods. It comes in two isomeric versions: alpha-Citral (or geraniol), which has a strong lemon fragrance, and beta-Citral (or neral), which has a softer lemon fragrance.
+ **Curcumene:** Directly related to turmeric.
+ **Eucalyptol (or cineol):** A monoterpene that is the major compound in eucalyptus, which has a cold taste and is also found in rosemary, sage, wormwood, and basil.
+ **Gingerol:** One of ginger's important phenolic compounds, with a biting taste that translates as a kind of heat. Chemically, gingerol is a close relative of capsaicin (see the chapter of the same name), the active component of hot peppers. The concentration of gingerol is high in fresh ginger and lower in dried ginger. On the other hand, once dried, gingerol diminishes in favor of shogaol, a compound that is twice as spicy as gingerol. When exposed to heat, gingerol degrades to zingerone, which has a sweeter flavor.
+ **Linalool:** A terpenic alcohol with the fragrance of lavender and lily of the valley, which also occurs in bergamot, rosewood, and mint.
+ **Nerolidol:** Also known as peruviol, this is one of the main volatile compounds of ginger, along with citral and beta-sesquiphellandrene. It adds woody touches and also occurs in beer, lemongrass, orange blossoms, strawberries, jasmine, lavender, and green tea.
+ **Paradol:** A phenolic compound with a peppery taste, chemically very close to zingerone. It is also found in Guinea pepper.

+ **Shogaol:** A compound found in dried ginger, twice as spicy as gingerol.
+ **Zingerone:** When ginger is subjected to heat, gingerol degrades to zingerone, which is much sweeter than gingerol.
+ **Zingiberene:** A compound also found in turmeric.
+ **As well as several other molecules:** Alpha-pinene, limonene, borneol, farnesene, geraniol, paracymene, and myrcene.

## FRESH OR COOKED GINGER?

Cooked ginger tastes sweeter than raw ginger, but if it is added towards the end of cooking or directly to the dish at serving time, it is more biting! Be sure to take this into consideration when selecting a wine to accompany a dish spiced with ginger.

## COMPLEMENTARY INGREDIENTS FOR COOKING WITH GINGER

There are a number of complementary ingredients that possess a molecular structure similar to that of ginger. We can use them alongside ginger to solidify the harmonic bridge between the flavors on the plate, or to combine dishes more

## 2. PRINCIPAL COMPLEMENTARY FOODS
### GINGER

| | | |
|---|---|---|
| BEER | GUINEA PEPPERS | ORANGES |
| BERGAMOT | HOT PEPPERS | PARSLEY |
| BORONIA FLOWERS | JASMINE | PIMENTÓN |
| CARDAMOM | LAVENDER | ROSEMARY |
| CRANBERRIES | LEMON | ROSEWATER |
| DRIED ROSEBUDS | LEMONGRASS | SAGE |
| EUCALYPTUS | LEMON VERBENA | STRAWBERRIES |
| FINGER LIME | LYCHEES | TOBACCO |
| FRESH FIGS | MANGOS | TURMERIC |
| GALANGAL | MATE | YUZU |
| GRAPEFRUIT | MINT | |
| GREEN TEA | ORANGE BLOSSOMS | |

## 3. SECONDARY COMPLEMENTARY FOODS
### GINGER

| | | |
|---|---|---|
| BALSAMIC VINEGAR | HARD CHEESES FROM SUMMER PASTURES | STAR ANISE |
| CINNAMON | RASPBERRIES | SWEET EUROPEAN BASIL |
| CORIANDER SEEDS | | VANILLIN |

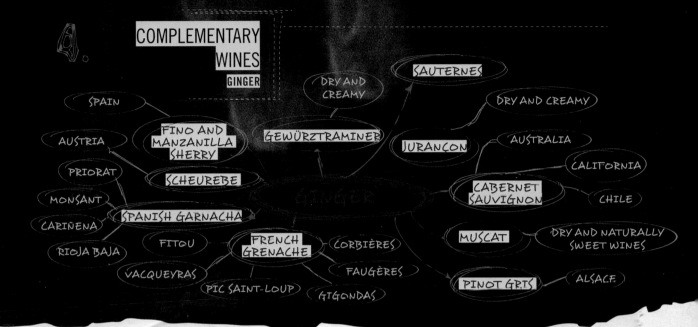

# COMPLEMENTARY WINES
## GINGER

- SPAIN
- AUSTRIA
- PRIORAT
- MONSANT
- CARIÑENA
- RIOJA BAJA

**FINO AND MANZANILLA SHERRY**

**SCHEUREBE**

**SPANISH GARNACHA**

- FITOU
- VACQUEYRAS

**FRENCH GRENACHE**
- PIC SAINT-LOUP
- GIGONDAS
- CORBIÈRES
- FAUGÈRES

DRY AND CREAMY

**GEWÜRZTRAMINER**

**SAUTERNES**

**JURANÇON**

DRY AND CREAMY

- AUSTRALIA
- CALIFORNIA
- CHILE

**CABERNET SAUVIGNON**

**MUSCAT** — DRY AND NATURALLY SWEET WINES

**PINOT GRIS** — ALSACE

GINGER

---

successfully with wines selected to accompany ginger (see Figure 2).

## THE TASTE OF COLD IN GINGER

Ginger is one of the group of cold-tasting foods (see the chapter "A Taste of Cold"), which I have so named because of the presence of aromatic compounds such as camphene and eucalyptol—in the case of ginger—that project the refreshing and gripping taste of camphor and eucalyptus.

The cold-tasting phenomenon is the same, for example, for estragole in apples and menthol in mint. These molecules activate taste receptors for temperatures between 8°C and 28°C (about 46°F to 82°F) and thus simulate cold—in contrast to the capsaicin contained in hot peppers, which increases the temperature of the taste buds.

This explains the sensation of coolness in the mouth, especially when ginger is savored raw—just as with apples, mint, and other cold-tasting foods (see the following list) when consumed in their natural state.

## "COLD-TASTING" INGREDIENTS COMPLEMENTARY TO GINGER

Apples, coriander, cucumbers, eucalyptus, fennel, fresh tarragon, green basil, green peppers, horseradish, lemon balm, lemongrass, limes, mint, parsley and parsley root, parsnips, raw celery and celery salt, verbena, wasabi, wild basil, and yellow carrots.

## WINES COMPLEMENTING GINGER'S COLD TASTE

Ginger's cold-tasting aromatic compounds, like the ingredients with the same refreshing molecules, augment the perception of cold, and thus can render a wine even colder!

These foods reinforce the perception of acidity and bitterness in wine. In effect, the physiognomy of taste explains that coldness increases the perception of acidity and bitterness, whether in a food or a wine, or during the meeting of the two.

So it is important to serve white wine at a higher temperature than usual with these foods, and to avoid white wines that have a biting acidity or are excessively bitter—unless, of course, you love bitter tastes, which can be very pleasant for those who are so inclined (see the "A Taste of Cold" chapter).

## A BIT OF MIXOLOGY

Ferran Adrià of elBulli was one of the first great chefs to redefine, at the beginning of the 1990s, aperitif cocktails with new techniques and new flavors. He revisited classic cocktails such as the Caipirinha, the Piña Colada, and the Kir Royal, among others.

Thus, a new bar/restaurant discipline was born: mixology. The bartender is now a mixologist.

Following is one of the variations that I amuse myself with when playing mixologist at cocktail hour.

### GINGER BELLINI (FRANÇOIS CHARTIER VARIATION):

This is a new twist on the Bellini, a cocktail traditionally made from peach purée and sparkling wine, that takes advantage of the affinity between beer and ginger.

**Primo:** Crown the edge of a large glass with crystallized ginger salt (it's sufficient to reduce it to fine grains). You may also use powdered ginger.

**Secundo:** Add, according to taste, fresh shredded ginger to mango nectar (about ½ to ⅔ of a glass) along with a few drops of yuzu juice (the yuzu is an East Asian citrus fruit). These are two complementary foods from the same molecular family as ginger that bring exotic flavors to the mix. The yuzu also adds a good acidity to the beer, which naturally has little acid.

**Tertio:** Complete with a blond beer (according to taste, ½ to ⅓ of a glass, with ½ to ⅔ of the glass containing the mango nectar/yuzu/ginger). Serve very cold.

Dare to try a variation on this cocktail by replacing the mango, yuzu, and ginger with cranberries, lemongrass, and galangal, all of which are from the same molecular family as ginger. You may also replace the beer with a sparkling wine. It's your call!

## RE-GINGERING...

In order to complete your understanding of ginger and its power of attraction over wines, continue the adventure by revisiting the "Gewürztraminer/Ginger/Lychee/Scheurebe" chapter.

GINGER BELLINI
FRANÇOIS CHARTIER VARIATION

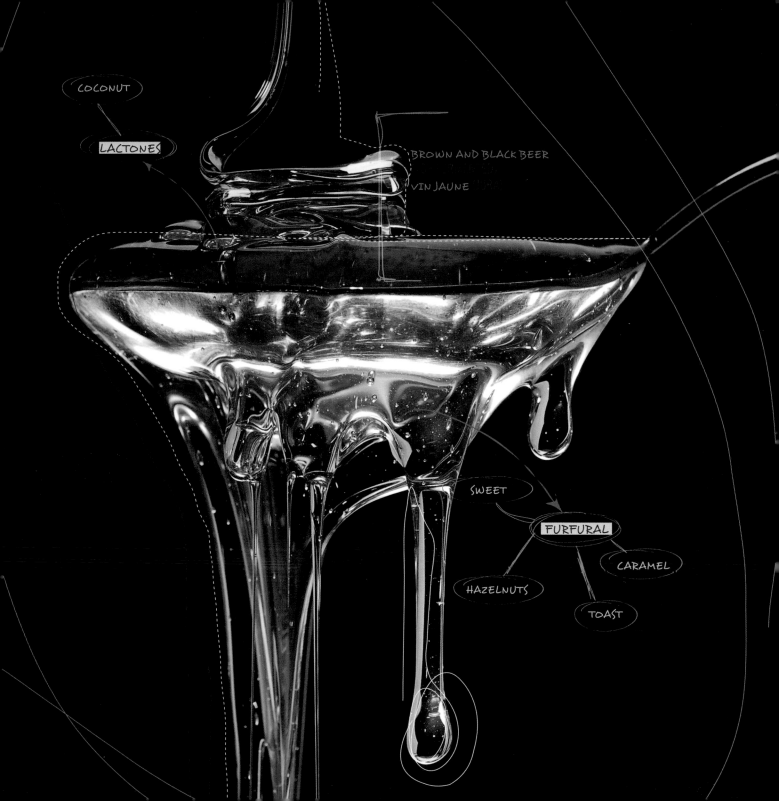

COCONUT

LACTONES

BROWN AND BLACK BEER

VIN JAUNE

SWEET

FURFURAL

CARAMEL

HAZELNUTS

TOAST

# MAPLE SYRUP

## THE AROMATIC SAP STAMPED WITH A QUEBEC IDENTITY

"We must be able to experience everything that is real,
and everything that we experience must be real."

WILLIAM JAMES

In this chapter, I'll share with you the results of my research on food and wine harmonies as it relates to the volatile compounds of the maple tree and its precious syrup.

We'll follow the path of the enchanting fragrances of maple syrup, and of the wines and other beverages that find a perfect match in foods featuring maple products.

### THE HISTORY, PRODUCTION, AND QUALITY OF MAPLE SYRUP

Originating in China and Japan, the maple family consists of more than 125 varieties, of which four North American species, including the sugar maple, are cultivated for their maple syrup. *Acer saccharum* (sugar maple), a species that grows primarily in northeastern North America, is responsible for most maple syrup production.

More than half of the world's maple groves are found in the United States, but 75% of the syrup enjoyed across the globe comes from the 7,400 maple sugar growers of Quebec.

A cold winter, a thick layer of snow on the ground, major differences between daytime and nighttime temperatures during the first days of spring, and a lot of sunshine are the *sine qua non* conditions for producing excellent maple syrup.

During the course of the approximately six-week-long harvest season, a tree provides about 5 to 20 liters (approximately 6.5 to 26.5 gallons) of maple water. (Some trees can provide up to 320 liters, or about 440 gallons!) It takes about 40 liters (55 gallons) of maple water to produce one liter (about $1\frac{1}{3}$ gallons) of syrup.

At the beginning of the season, the sap contains about 3% sucrose, an amount that is halved by the season's end. So as spring progresses, producers must heat the liquid for a longer period of time. This means a darker and (depending on one's tastes) tastier syrup. The longer the maple water is boiled, the darker and thicker the syrup will be.

Today, many maple syrup producers use the inverse osmosis process, as do many wine producers. In the case of maple syrup, this process is used to remove 75% of the water in the sap without needing to heat it. Producers then boil the remaining concentrated sap, increasing its sugar content.

The final composition of the syrup is about 62% sucrose, 34% water, 3% glucose, and 0.5% malic acid and other acids, as well as some traces of amino acids, which furnish the maple syrup with a fullness and presence in the mouth.

According to the Canadian grading system, Grade A syrup is clear, with a delicate flavor, meant to be savored alone or used as a condiment. Grades B and C are syrups more richly endowed with caramelized flavors—mostly coming from furanone, maltol, cyclotene, and sotolon, the dominant aromatic molecules—and so are generally used in cooking certain traditional dishes and to glaze meat and fish.

## MAPLE SYRUP'S TASTE AND FRAGRANCE

Maple syrup's taste is dominated primarily by sugars and by a great complexity of flavors that lasts a long time in the mouth. Also present, but more discreet because of their domination by the syrup's imposing sweetness, are the bitter flavors that come from some phenolic acids.

As for maple syrup's fragrance, that is characterized by the dominant toasted aroma of roasted fenugreek seeds and the scent of vanilla, as well as by caramelized and empyreumatic tones resulting from the carmelization and the browning reaction between sugars and amino acids (Maillard reaction) during the cooking of the maple water. It takes just one whiff of roasted fenugreek seeds to realize the astonishing aromatic parallel between this Indian spice and maple syrup.

The next time you have maple syrup under your nose, just remember that we can detect, from one syrup to another, more than 250 aromatic references coming from about a hundred volatile compounds!

The aromas detected in maple syrup can be grouped into different families associated with vegetal, floral, fruity, spicy, milky, and even candy-like empyreumatic aromas.

### THE MAJOR AROMAS FOUND IN MAPLE SYRUP INCLUDE:

Almonds, brown sugar, burnt wood, butter, cedar, chicory, chocolate, cinnamon, cloves, coffee, cream, cut hay, fir trees, flowers, golden sugar, hazelnuts, honey, licorice, marshmallows, milk, molasses, mushrooms, oats, roasted fenugreek seeds, rye, sponge toffee, toast, vanilla, walnuts, and wheat.

## THE MOLECULAR STRUCTURE OF MAPLE SYRUP

In the complex fragrance of maple syrup, one aromatic molecule in particular is dominant: maple furanone (5-ethyl-3-hydroxy-4-methyl-5H-furan-2-one), also known as ethyl fenugreek lactone. This is a lactone even more powerful than sotolon, with a caramelized aroma of maple and roasted fenugreek seeds.

Remember: maple syrup and roasted fenugreek seeds are almost twins!

The maple furanone found in maple syrup occurs naturally in soy sauce, providing its unique bouquet—which scientifically explains the frequent pairing of soy sauce and maple syrup in the glazing of fish and meat.

We also find in maple syrup another furanone, 2-5-dimethyl-4-hydroxy-3(2H)-furanone, an active compound also called strawberry furanone, which has the caramelized and fruity fragrance of burnt pineapple and cooked strawberries. This compound is also present in roasted hazelnuts and almonds, coffee, soy sauce, popcorn, cooked tacos, malt, beer, Swiss cheese, and boiled beef. These are all complementary foods for cooking with maple syrup.

Other lactones are found in maple syrup as well. These aromatic molecules, also typical of wines raised in oak barrels, have tones of fruit (apricots/peaches and coconut), nuts (almonds and hazelnuts), or caramel.

Among the numerous aromatic compounds found in maple syrup is maltol, with the fragrance of burnt sugar, playing in the fruity/sugary sphere of cotton candy (note that maltol is used as a flavor enhancer for cotton candy, as well as for many other products), also typical of the woody and torrified notes of wine raised in oak barrels. Maltol is also found in Madeira, tawny port, chicory, cocoa, coffee, cooked milk, roasted malt, cooked strawberries, and bread crusts.

### BURNT SUGAR: MULTIPLE AROMATIC ORIGINS...

Numerous active molecules can be responsible for the typical burnt sugar fragrance associated with certain foods, either cooked (caramel, cotton candy, maple syrup, Madeira, strawberries, and pineapple) or roasted (chicory and coffee). These include: furanone acetate, coronol, methyl coronol, cyclotene, ethyl cyclotene, furanone, maltol, ethyl maltol, maple furanone, mesifurane, and sotolon.

Maple syrup also assumes part of its aromatic identity from a molecule called sotolon (see the chapter of that name), whose complex fragrance is reminiscent of walnuts, curry, fenugreek seeds, caramel, and maple. This important compound, which has a powerful aroma, also helps define the aromatic identity of curry, fenugreek seeds, soy sauce, balsamic vinegar, coffee, cotton candy, cooked celery, and celery salt.

Sotolon also plays a part in the aromatic profile of dark beer, vin jaune from Jura, Montilla-Moriles wines, sherries (especially amontillado and oloroso), tawny port, Madeira, red and white VDN wines (raised in an oxidative milieu), aged Champagne, Vino Santo, and certain sakes, as well as aged brown rum and sweet wines made from grapes affected by noble rot *(botrytis cinerea)*, as are Sauternes and Tokaji wines (ideally about twelve years old).

The long list of aromatic compounds that intermingle to produce maple syrup's fragrance also includes cyclotene (methyl cyclopentenolone). Cyclotene is found in all grilled or roasted foods that contain sugars. It resembles furanone and maltol, giving off a very powerful fragrance halfway between licorice and maple syrup.

Cyclotene naturally occurs in roasted almonds, coffee, cocoa, and roasted fenugreek seeds, and to a lesser extent in black currants, onions, wheat bread, dried and cooked pork, beer, barley, sukiyaki sauce, licorice, dried bonito tuna, and roasted chicory roots.

In coffee, ethyl cyclotene acts as a flavor enhancer. So, if you cook with both maple syrup, already rich in cyclotene, and coffee, which contains ethyl cyclotene, the coffee will have the effect of giving more presence and amplitude to the flavors of maple syrup.

### COFFEE: A FLAVOR ENHANCER

Try pairing coffee with other foods and drinks that act as flavor enhancers, such as maltol, sherry, and asparagus, as well as umami-tasting foods (rich in glutamate). Take advantage of these ingredients to enhance your recipes!

It's impossible to discuss maple syrup without mentioning eugenol, an important woody, spicy molecule that is the main active component in the fragrance of cloves (see the chapter of that name) and in the fragrance of wines raised in oak barrels.

ETHYL CYCLOTENE = FLAVOR ENHANCER

1.

SAME VOLATILE
COMPOUNDS
**MAPLE SYRUP AND WINES
RAISED IN OAK BARRELS**

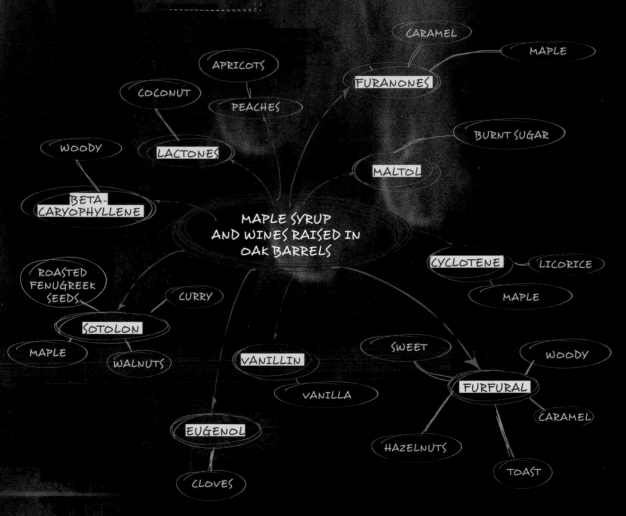

MAPLE SYRUP
AND WINES RAISED IN
OAK BARRELS

CARAMEL

MAPLE

FURANONES

APRICOTS

COCONUT

PEACHES

BURNT SUGAR

WOODY

LACTONES

MALTOL

BETA-
CARYOPHYLLENE

CYCLOTENE — LICORICE

MAPLE

ROASTED
FENUGREEK
SEEDS

CURRY

SWEET

WOODY

SOTOLON

FURFURAL

MAPLE

WALNUTS

VANILLIN

CARAMEL

VANILLA

HAZELNUTS

TOAST

EUGENOL

CLOVES

## 2. PRINCIPAL COMPLEMENTARY FOODS
### MAPLE SYRUP

APRICOTS
BALSAMIC VINEGAR
CHERRIES
CINNAMON
CLOVES
COCOA/DARK CHOCOLATE
COCONUT
COFFEE
COOKED CELERY AND CELERY SALT
COOKED MILK
COOKED TACOS
COTTON CANDY
CRUSTACEANS
CURRY

EUCALYPTUS
GRAPES
GRILLED ALMONDS
GRILLED, SMOKED, OR ROASTED MEAT
LICORICE
MARSHMALLOWS
MUSHROOMS
PEACHES
PINEAPPLE
PLUMS
POPCORN
POWDERED MALT
RASPBERRIES
ROASTED CHICORY

ROASTED FENUGREEK SEEDS
ROASTED HAZELNUTS
ROASTED PEANUTS
SANDALWOOD
SMOKED FISH
SOY SAUCE
STRAWBERRIES
SWISS CHEESE
TOAST
TOBACCO
TONKA BEANS
VANILLA
YLANG-YLANG

## 3. COMPLEMENTARY WINES AND BEVERAGES
### MAPLE SYRUP

BROWN AND BLACK BEER (TOP FERMENTED)
VIN JAUNE
MONTILLA-MORILES (AMONTILLADO AND OLOROSO)
MADEIRA
TAWNY PORT
SHERRY
NIGORI SAKE

BROWN RUM
AMERICAN BOURBON
BRANDY
KIRSCH
AMARETTO

To the long list of molecules found in maple syrup, we can also add benzaldehyde, an important aromatic molecule in almonds, apricots, cherries, strawberries, raspberries, peaches, plums, and grapes, which also provides fragrance to kirsch and amaretto; furfural; beta-caryophyllene, with its woody notes; as well as numerous terpenes, which provide fragrance to conifers and citrus fruits. You probably never imagined that our traditional maple syrup was so complex!

## OAK BARRELS AND MAPLE SYRUP: THE SAME PROFILE...

When we examine the list of maple syrup's active compounds, we note their very great similarity with those compounds found in wines raised in oak barrels. (You may recall this from the "Oak and Barrels" chapter.)

This is because oak and maple are two essences that, when transformed by heat—the oak barrels' insides are charred prior to use, and maple water is heated at a high temperature to transform it into syrup—see their active compounds transform into an assortment of new molecules that are even more complex and aromatic.

Another similarity, which I have touched on elsewhere, is that wines raised in oak barrels and maple syrup both contain lactones, which have notes of fruit (apricot/peach), nuts (almonds and hazelnuts), and caramel. Included among these are furanones, particularly maple furanone, which smells of caramelized maple.

Oak barrel–raised wines and maple syrup also have in common maltol (burnt sugar), cyclotene (a strong fragrance situated between licorice and maple), furfural (whose fragrance is of sugar/wood/toast/hazelnut/caramel, with burnt touches), vanillin (vanilla), eugenol (cloves), sotolon (a complex fragrance associated with walnuts/curry/fenugreek/caramel/maple syrup), beta-caryophyllene (woody fragrances), and many other molecules shared by oak and maple products.

All of these common compounds and fragrances mean that maple syrup–sweetened dishes and wines raised for a long time in oak barrels and casks—such as Sauternes, tawny ports, Madeiras, and oloroso sherries—pair very harmoniously.

We must also not forget dry wines, both white and red, that carry the aromatic stamp of torrified barrels. These open the door to further harmonic pairings with savory foods that have been enhanced by the fragrances of maple syrup.

Furthermore, you may recall that wines raised in American oak barrels are more richly endowed with aromatic molecules closely related to those of maple, which means that these wines are even more closely linked to maple syrup. This is the case for some New World Chardonnays, and for some other California and Spanish red wines, such as those from Rioja and Ribera del Duero, in which American oak is still widely used.

Finally, brandies and whiskies that have spent a long time maturing in oak barrels (including cognac, Armagnac, American bourbon, and Scotch) can be considered good candidates for pairing with maple syrup–sweetened dishes.

## PHENOLIC COMPOUNDS

Wines and numerous foods, including vanilla, maple water, and especially the syrup that results from cooking this water, are very rich in diverse phenolic compounds. These include p-coumaric acid, a derivative of cinnamic acid found in peanuts, cinnamon, tomatoes, carrots, and garlic; and ferulic acid, a component of lignin that is found in wood, derived from cinnamic acid, and is a precursor of other aromatic molecules, such as vanillin.

Ferulic acid is also found in coffee, apples, artichokes, peanuts, oranges, pineapples, and 5-hydroxymethyl-2-furaldehyde, which is a significant presence in the fragrances of caramel and honey.

Other phenolic compounds found in maple syrup and some of its complementary foods include: homovanillic acid, protocatechuic acid, sinapinic acid, vanillic acid, coniferyl alcohol, catechin, coniferaldehyde, flavanol and dihydroflavanol derivatives, and syringaldehyde.

## THE WHEEL OF MAPLE SYRUP FLAVORS

If today we are more aware of maple syrup's aromatic structure, this is largely thanks to Jacinthe Fortin and her research team. Ms. Fortin is a sensory evaluation analyst with Agriculture and Agri-Food Canada's Food Research and Development Centre in Saint-Hyacinthe, Quebec. The results of her research on maple syrup can be found in the report "Flavour Wheel for

Maple Products," which was created by a team of seasoned analysts. (It is available without cost on the Agriculture and Agri-Food Canada website, www.agr.gc.ca.)

## AT THE TABLE WITH MAPLE SYRUP

As you may have noticed, the list of compounds that contribute to the complexity of maple syrup's bouquet and flavors is very long. The list of complementary ingredients that contain the same molecules as maple is even longer! The result is an abundance of possibilities for harmonious pairings, both among ingredients on the plate and between plate and glass.

## WINES FOR SUCCESSFUL PAIRINGS

The molecular complementarity between maple syrup and wines is fairly rich, thus enabling many lovely harmonic pairings.

Among dry wines, to accompany food that is more savory than sweet, there are New World Chardonnays and some red wines from California and Spain, in particular the regions of Rioja and Ribera del Duero (this, as we have seen, is because these are regions in which American oak is still quite present). Another good option is aged Champagne.

Among sweet wines, to accompany sweet or sweet/savory dishes, go for Sauternes and Tokaji Aszú (ideally about ten years old), naturally sweet red and white wines raised in an oxidative milieu, and Vino Santo.

Since some of maple syrup's aromatic molecules are also strongly present in cinnamon, it is a good idea to choose wines that go well with this spice. The best choice is ice cider from Quebec, followed by Gewürztraminer (late-harvest or SGN wines, made from grapes affected by noble rot), Grenache, and Pinot Noir.

We can also add to this list wines and beverages already mentioned for cooking with maple syrup: top fermented brown and black beers, vin jaune from Jura, Montilla-Moriles wines, sherries (amontillado and oloroso), Madeira (Bual and Malmsey), tawny port, sake, aged brown rum, American bourbon, brandies raised in oak, kirsch, and amaretto.

## HARMONIOUS PAIRINGS WITH MAPLE SYRUP

Everything we have learned so far in this chapter is meant to open up routes to new recipe ideas for cooking with maple syrup, as well as to present new wine pairing options for dishes flavored with this delicious product.

For example, if you were to cook a salmon with maple syrup and black beer, the sweetness of the syrup and the bitterness of the beer would be your key elements in successfully pairing this dish with a low-alcohol sake rich in amino acids, such as Gekkeikan Nigori Sake. With only a 10% alcohol content, served cold, it is creamy, quasi-sweet, and quite enveloping—and it pairs brilliantly with most maple syrup–sweetened foods.

## WHAT TO DRINK IN THE SUGAR HOUSE?

Among other options, two excellent choices would be a low-alcohol sake that is rich in amino acids, such as the unique, delectable, milky Nigori sake, or else a more classical sweet wine lightly touched by noble rot, ideally one that has spent several years in the bottle, such as some French vintages from Monbazillac, Sainte-Croix-du-Mont, and Sauternes.

You may be surprised to learn that when it comes to food pairing, some high-quality beers can be just as complex as wine. For example, in a flared wine glass, a tad cooler than room temperature, serve a black beer, such as the penetrating, chocolatey, roasted-flavored Boréale Noire Stout, Les Brasseurs du Nord, Québec (5.5% alcohol), which can really hold its own with the sweet/savory flavors of maple-glazed ham.

There are similar considerations for selecting a wine to accompany maple-glazed pork medallions with sweet potatoes, garnished with spicy pecans. Between the marinade dominated by the maple syrup's sugar and the fruity acidity of cider vinegar, as well as the subtle sugary tones of pecans and sweet potatoes, there is really no other choice: we must leave the beaten path. Dare to try an eclectic vin jaune from Jura, such as those from the houses of Rolet and Stéphane Tissot, with aromas of walnuts, curry, and maple. These are very dry, minerally, and bracing in the mouth. These wines are sensational even against the imposing saltiness of deep-fried smoked pork jowls, a specialty of Quebec!

You may also choose to follow a sweeter path by serving a caramelized ten-year-old tawny port at a cool temperature or a sweet wine lightly affected by noble rot, such as Sauternes or vintages from neighboring areas. All of these made-to-measure companions work brilliantly with maple syrup and leave behind the traditional, less inspired glass of milk…

## SOME WINE AND FOOD PAIRING CREATIONS CONCEIVED FOR SPECIAL EVENTS

In the chapters on sotolon and cloves, I shared with you parts of the menu for "The Spice Route," the meal that I created in conjunction with chef Racha Bassoul at his former Montreal restaurant, Anise, in 2007.

Now that we know a bit more about the aromatic structure of maple syrup, I present here again the second act of the meal, which was centered around sotolon and maple syrup:

### Cuvée Sotolon —
### Inspired by the Romans
### and conceived by François Chartier

*Based on a young, dry Jurançon, in which roasted
fenugreek seeds have been macerated
to provide a more mature, spiced profile
in the image of classic Roman wine*

### and
### Three Princesses and Three *Espumas*

*Princess scallops and their coral, and three types of
espumas (sake and salt water; roasted fenugreek seeds
macerated in wine; maple syrup), accompanied
by watercress and red shiso.*

This course was designed with harmony in mind, with the ingredients and the wine united by a shared aromatic molecule: sotolon. There was iodine in the scallops, the sake, the maple syrup, the roasted fenugreek seeds, and the seawater of the *espumas* (airy mousses resembling foam, produced by a siphon, without any fats).

Everything in the dish and in the wine was inspired by sotolon, dominating in the powerful aroma of the roasted fenugreek seeds and echoed in the aromatic signature of maple syrup.

## SOME OTHER PAIRING IDEAS CONCEIVED FOR SPECIAL EVENTS

+ Maple-glazed salmon in a soy-balsamic vinegar reduction with sautéed mushrooms **Asuncion Oloroso, Montilla-Moriles, Alvear, Spain**
+ Catalan/Indian/Quebecois surf and turf with fried foie gras and well-grilled scallops in a curry-maple reduction (a variation of the recipe for "Surf 'n' turf anise" in *À table avec François Chartier*) **Vin jaune 1985 Château-Chalon, Henri Maire, France**
+ Croustade of foie gras with apples, gingerbread, maple syrup, and curry (a variation of the recipe for "Croustade of foie gras with apples" in *À table avec François Chartier*) **Tokaji Aszú 5 Puttonyos, Disznoko, Hungary**

GREEN BASIL
CARROTS
QUINCES
CORIANDER
TURMERIC

EXTRA-STRONG
BEER

OAKED CHARDONNAY

GALANGAL
GINGER
ROASTED SESAME SEEDS
DRIED AND AGED HAM  (SPANISH HAM,
PROSCIUTTO)

SEMI-FIRM AND
HARD CHEESES

# QUEBECOIS AND CLASSIC EUROPEAN CHEESES

## TRACKING DOWN THEIR AROMAS

"From my earliest years,
I have taken a lot of incorrect opinions to be true."

RENÉ DESCARTES

I am now going to take you into the heart of my work detecting the volatile molecules associated with the complex flavors of the greatest Quebec cheeses, as well as the classic Europeans. This, of course, includes the wines that harmonize so well with them.

This chapter is divided into three distinct themes: semi-firm and hard cheeses (Theme I), bloomy-rind cheeses (Theme II), and blue cheeses (Theme III). But before we do anything else, let's take a look at cheese's general molecular structure.

### CHEESE'S AROMATIC STRUCTURE

Most cheeses contain, among other compounds, carotenoids, which give them their yellow color, as well as peptides and fatty acids. The latter act as flavor enhancers for sheep's milk and lamb's milk, and provoke a peppery sensation on the tongue when one is savoring blue cheese.

In addition to carotenoids, we also have methyl ketones, which give blue cheeses their characteristic flavor; numerous amino acids (part of the umami flavor), such as putrescine (tasting of gamy meat) and trimethylamine (smelling of fish, such as herring); and sulphuric and ammoniac compounds.

Also note the presence of diacetyl, whose strong, penetrating butter fragrance is recognized as one of the most important aromatic signatures of cheese and certain other dairy products. Diacetyl is found in many other products as well, including lavender oil, some flowers (narcissus and tulips), fino

sherry, white wines from hot countries raised in oak barrels on their lees, top fermented beer, coffee, tea, cognac, Scotch, cabbage, green peas, tomatoes, guava, honey, chicken, pork, and beef. These products thus become complementary ingredients and beverages for cooking and pairing with cheese.

Another component of cheese is acetoin, an active compound with a fatty, creamy, buttery taste reminiscent of butter and yogurt. Acetoin is also an important volatile compound of fino and manzanilla sherries (see the "Fino and Oloroso" chapter), and a major factor in providing the aromatic identity of cheese and a number of other foods and beverages, as you will see below.

### FOODS AND BEVERAGES BEARING THE MARK OF ACETOIN

Asparagus, broccoli, Brussels sprouts, butter, cantaloupe, corn syrup, fermented tea, milk, quinces, raw and cooked apples, raw and cooked leeks, roasted coffee beans, strawberries, top fermented beer, white wines (hot-country, raised on their lees in oak barrels), and yogurt. These ingredients all have a high degree of molecular compatibility, and when combined, they open the door to a whole world of possibilities for dishes that feature cheese, and for successful pairings with fino and manzanilla sherries and other wines.

Note that the aromatic tonalities of toast and crackers that can be detected in some cheeses are generated by two volatile

CORIANDER

RIESLING

TURMERIC

molecules, cyclotene and maltol. In wine, these two aromatic compounds come from oak barrels that were charred before use. This supports the pairing of cheese with wines raised in barrels, particularly those based on Chardonnay. Cyclotene and maltol also contribute to the bread and cracker fragrances found in beer, toast, and pastries.

## THEME I
## SEMI-FIRM AND HARD CHEESES

A sizeable proportion of Quebec's best farm-raised cheeses, in particular those sold in the autumn and the beginning of winter, come from summer pastures. This means that they were made from the milk of herds that grazed in fields brimming with fresh herbs and flowers, in July and August. The result is very fragrant cheeses, with complex aromatic notes that can be attributed to the animals' fresh, natural, and often organic diet.

Semi-firm and hard cheeses, ripened for two to three months, express themselves at the heart of the cheesemaking season with herbal and flowery tones caused by aromatic molecules from the terpene family, such as the very fragrant linalool, which has the fragrance of lavender and lilies of the valley.

These cheeses often have a yellow color, more or less pronounced depending on the cheese, that originates in carotenoid pigments. This brings to mind the molecular structure of saffron, which as you may recall is also rich in linalool (flowers) and carotenoids. Linalool also has a strong presence in fino sherry, as well as in Muscat, Gewürztraminer, and Riesling white wines. The latter grape variety has the most forceful notes of terpenes and carotenoids, the two principal aromatic molecule groups characterizing the flavor of semi-firm and hard cheeses.

Carotenoids can impart various colors to foods. They give a yellow or orange color to many fruits, vegetables, and spices, including carrots, quinces, pears, apples, and saffron, and they give a red color to tomatoes and watermelons.

In wine, carotenoids are especially present in "non-aromatic" grape varieties, such as Chardonnay, Sauvignon Blanc, and Riesling, as well as in some rosé wines.

This scientifically confirms what many sommeliers know from experience: the privileged place that white wine holds with cheeses.

With semi-firm and hard cheeses, go for Rieslings with a flowery fragrance, dry or sweet Muscats and Gewürztraminers, fino and manzanilla sherries, certain non-oaked, floral Chardonnays, or rosés.

Finally, given that floral fragrances engendered by linalool are also found in green basil, coriander, and lychees, and that carotenoids can also be found in carrots, quinces, pears, apples, and saffron, dare to cook or accompany your cheeses with these ingredients. They will enhance your dining experience and strengthen the harmonious links with your selected wine.

### SOMMELIER-COOK'S HINT

Coat a semi-firm or hard cheese with coriander seeds, or accompany the cheese with a mesclun salad spiced with fresh coriander or basil, accompanied by several apple or pear quarters. Serve with a white wine such as Riesling. Why not accompany this cheese with a saffroned julienne of carrots (lightly cooked, which allows them to develop a floral note of violets)?

**SOME PAIRINGS TO EXPLORE WITH QUEBEC CHEESES**

Serve a young, very floral Alsatian Riesling, such as Domaine Fernand Engel Riesling Réserve, Alsace, France, with a cheese plate featuring two fine Quebec cheeses. Le D'Iberville from Fromagerie Au Gré des Champs (Saint-Jean-sur-Richelieu) and Le Grondines from Fromagerie des Grondines (Portneuf). The cheeses' herbaceous and floral fragrances and milky, salty flavors will find an echo in this ample and vivacious wine.

Should you choose to go with more substantial cheeses, such as Le Pied-de-Vent from Fromagerie du Pied-de-Vent (Havre-aux-Maisons, Îles-de-la-Madeleine) and Le Gré des Champs from Fromagerie Au Gré des Champs (Saint-Jean-sur-Richelieu), choose a white wine that's a bit denser and richer, with similar fragrances, such as the remarkable Charles Hours Cuvée Marie Jurançon Sec 2006, France. This fine, modestly priced white wine should be carafed and served at 14°C to 15°C (about 57°F to 59°F) to allow it to reach its true potential and benefit the most from its pairing with these two cheeses.

1.

**THEME 1**

# COMPLEMENTARY FOODS
**SEMI-FIRM AND HARD CHEESES**

APPLES
CARROTS
CORIANDER
GALANGAL
GINGER

GREEN BASIL
LAVENDER
LYCHEES
MUSCAT GRAPES
PEARS

QUINCES
SAFFRON
TURMERIC
VIOLETS

**THEME 1**

# COMPLEMENTARY WINES
**SEMI-FIRM AND HARD CHEESES**

NON-OAKED

CHARDONNAY

FLORAL

RIESLING

SAUVIGNON BLANC

FINO SHERRY

BEER

WHITE

SEMI-FIRM AND
HARD CHEESES

MANZANILLA SHERRY

MUSCAT

ROSÉ WINE

DRY OR SWEET

GEWÜRZTRAMINER

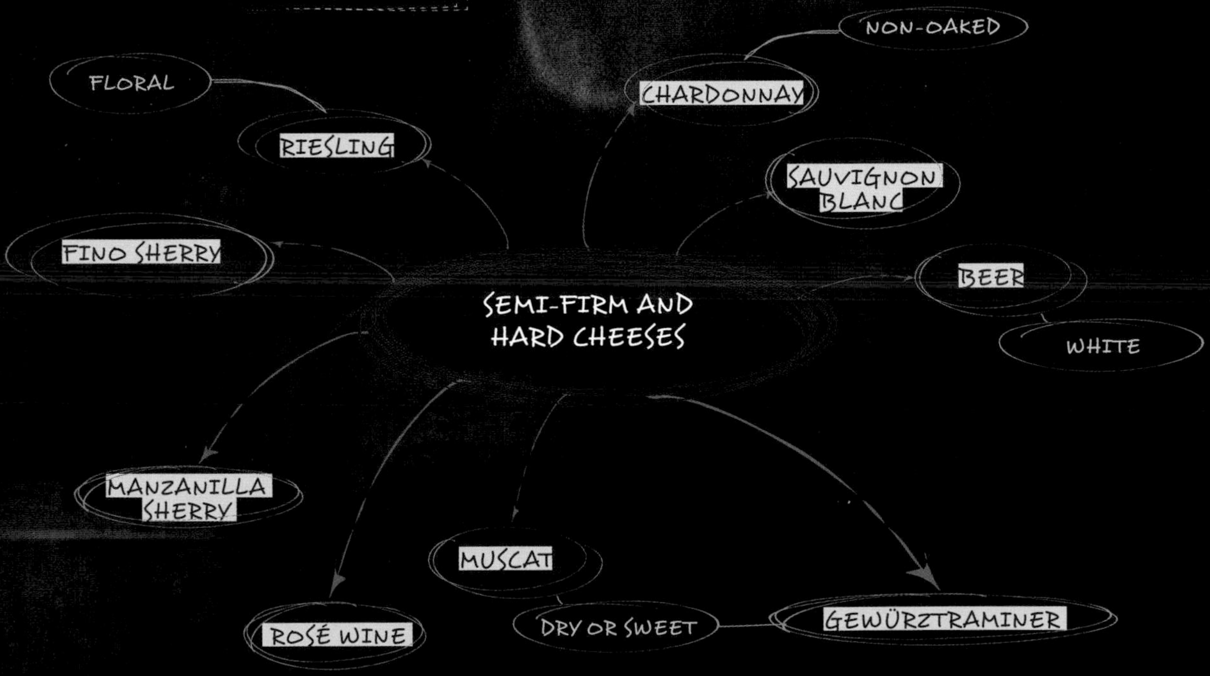

When it comes to cheeses that have ripened for a longer period of time, such as Le Valbert, aged 90 to 180 days, from Fromagerie Lehmann (Hébertville, Saguenay-Lac-Saint-Jean), and Alfred le Fermier, aged six to eight months, from Fromagerie La Station (Compton), allow their flavors of herb and dried flowers to fully express themselves by pairing them with a sherry. One good choice is the dry, compact, and fragrant Manzanilla Papirusa, Emilio Lustau, Spain.

Finally, if you like to finish a meal by accompanying your cheeses with a port or a VDN wine, select Le Cru des Érables from Les Fromages de l'Érablière (Mont-Laurier). This soft washed-rind cheese has ripened for sixty days in the cellars of the proprietors' family sugar house. It is aged in Charles-Aimé Robert, an "acéritif," a Quebecois port-like aperitif made from distilled maple sap (so named for *acer*, the Latin word for maple).

The complex, penetrating flavors of Le Cru des Érables are dominated by the maple tonalities of the sotolon family. As we saw in the chapter dedicated to sotolon, this aromatic molecule also plays an important part in the aromatic profile of curry, walnuts, and roasted fenugreek seeds, as well as of red and white VDN wines (raised in a oxidative milieu), Sauternes that are at least ten years old, aged Champagnes, aged rum, tawny port, Madeira, and some oloroso sherries and Montilla-Moriles wines.

All of these foods and wines, therefore, pair very well with the unique Le Cru des Érables. Another good option is the delicious ten year old Warre's Otima Tawny Port, Warre & Co., Portugal. With its inviting nose of brown sugar, maple, figs, walnuts, and spices, it is ample, smooth, and powerful in the mouth.

The excellent, subtle, and willowy VDN white wine, Cazes Rivesaltes Ambré 1995, France, which is made from the Grenache Blanc grape, is characterized like no other by fragrances from the sotolon family, and thus will be just as dazzling with Le Cru des Érables.

## SWISS, MIMOLETTE, AND PARMIGIANO-REGGIANO CHEESES

These cheeses are particularly characterized by dimethylpyrazines (with fragrances of coffee, cocoa, chicory, and maple syrup), as well as by fatty acids, including fruity esters (with pineapple and strawberry tones) and lactones (apricots/peaches/coconut). This explains their great ability to pair with a very wide variety of wines and with dishes based on these foods.

## THEME II
## BLOOMY-RIND CHEESES

Now we turn to the buttery and creamy flavors of Quebec cheeses with a bloomy rind (a velvety white surface mold, such as you see on Brie and Camembert), and, of course, the wines that pair perfectly with these phenomenal cheeses.

I have found that the more I delve into the molecular structure of different cheeses, the more I find confirmation of their impeccable harmony with white wines, which I have been suggesting for several years. So red wines don't have much of a place here, unless we fiddle with these bloomy-rind cheeses by stuffing them with condiments, such as a black olive paste for Syrah wine or ground cloves for vintages from Bierzo, Spain (you can find a recipe for this in the "Cloves" chapter).

A scientific understanding of wine and food harmonies can open the door to more judicious pairing of white wines and beers with different cheeses, and inspire new ideas for "denaturing" some cheeses to help them accompany red wines. This sharper understanding of cheese has enabled me to polish my pairing approach—first in theory, and then, returning to the glass and the plate, in practice.

Let's consider, for example, bloomy-rind cheeses such as Brie and Camembert, some of which are double-cream and others triple-cream. The flavors of these creamy cheeses are dominated by numerous volatile compounds, including some amino acids, which add volume and presence in the mouth (umami).

These cheeses also contain diacetyl, a cetone responsible for the characteristic smell of butter (and, to a large extent, of human sweat!). Diacetyl is present in dairy products such as butter, cream, and cheese, especially those with a bloomy rind.

BARREL-RAISED CHARDONNAY

UNCTUOUS WHITE WINES

BELGIAN BEERS
ENGLISH BEERS

But that's not all. Diacetyl is an aromatic molecule also found in wines, giving oaked Chardonnays raised on their lees their characteristic buttery note, in particular those from warm climates such as California and Australia. Indeed, raising wine in oak barrels is very popular in those regions. The higher the alcohol level of these wines, the more butter and its derivative flavors, such as caramel, are present. Raising wine in barrels *sur lie* also generates a good amount of amino acids, such as are also found in cheese.

The same phenomenon also occurs in top fermented beer with a fairly high level of alcohol, in particular English beers such as pale ale and brown ale, as well as dark Belgian beers. Diacetyl and amino acids, tasting of umami, play an important role in the flavors of these beers, and thus in their pairing with cheeses. That explains that!

### SOME PAIRINGS TO EXPLORE WITH QUEBEC CHEESES

The two Quebec cheeses most similar to Camembert are Casimir from Les Fromages de l'Érablière (Mont-Laurier) and Le Petit Normand from Fromagerie La Suisse Normande (Saint-Roch-de-l'Achigan). With these two cheeses, go for a very buttery, creamy, and enveloping New World Chardonnay, such as Le Bonheur Chardonnay, Simonsberg-Stellenbosch, South Africa. This is a typical southern Chardonnay, with

woody fragrances that aren't excessive, giving off notes of crème fraîche, vanilla, and sweet spices. It is round, fresh, and persistent in the mouth.

Champayeur cheese from the Fromagerie du Presbytère (Sainte-Élizabeth-de-Warwick) has a fairly compact and chalky bloomy rind and is not as runny as the other cheeses in its category, which resemble Camembert more. Here you will want a Chardonnay with a less "fatty" structure to harmonize with the cheese's compactness. A fine answer is the half-European, half-Australian approach of Scotchmans Hill Swan Bay Chardonnay, Victoria, Australia. This expressive white wine has moderate acidity, an ample body, and expressive flavours reminiscent of Golden Delicious apples, pineapples, and fresh butter.

### SOMMELIER-COOK'S HINT

**Bloomy-rind cheese tailored for a pairing with red wine** This is a simple, effective trick to help your favorite red wine accompany the type of cheese that usually has no mercy for the tannins in red wine: cut the cheese horizontally in two pieces. Sprinkle finely ground cloves on the bottom half. Cover with the top half, wrap, and macerate for several days at a cool temperature (see the picture in the "Cloves" chapter). This is a made-to-measure food pairing for ample-structured, barrel-aged red wines such as Spanish wines from Bierzo, Rioja, and

Ribera del Duero, as well as California Zinfandels. You may replace the cloves with a black olive paste to harmonize with a Syrah wine, or with a purée of sundried tomatoes for pairing with a New World Pinot Noir.

The famous Riopelle cheese from the Fromagerie de l'Île-aux-Grues (Île-aux-Grues), a very ample, captivating triple-cream, deserves a more complex and substantial white wine, such as Cloudy Bay Chardonnay, Marlborough, New Zealand. This is an engaging white, with elegant, expressive fragrances. It is dense, persistent, gripping, and refreshing in the mouth, leaving traces of pineapple, apples, roasted almonds, and crème fraîche. Its lovely harmony and complexity are reminiscent of some Côte-de-Beaune vintages.

## THEME III
## BLUE CHEESES
**"Cheese is milk made immortal."**
Ramón Gómez de la Serna

When we start to decipher the molecular structure of blue cheese, we find that certain volatile compounds dominate and thus play a large part in characterizing its unique taste.

In blue cheese, we find aromatic molecules including diacetyl, which smells of butter; methyl ketones, whose flavors are peppery, spicy, fruity, and meaty; the very fruity thiols; terpenes, such as the floral linalool; other ketones with effluents from controlled oxidation; acetone, reminding us of nail polish(!); as well as some dimethylpyrazines that give off odors of roasted coffee and cocoa.

We also find here fatty acids such as glutamate (a key part of umami, the fifth basic taste), which supports the ensemble, giving it lots of taste and expressivity in the mouth. It also has a flavor-enhancing effect on sheep's milk cheese, among other foods, with a peppery effect on the tongue.

The important presence of umami in blue cheeses means that when choosing a wine to accompany them, it is important to choose one with volume and a strong presence in the mouth. Good examples of this would be sweet wines, VDN wines, ports, Chardonnays from warm countries that have slowly matured on their lees in oak barrels, and top fermented beers.

These beers and Chardonnays also generate the buttery tones associated with diacetyl, another attribute that makes them a logical choice to accompany blue cheese. Note that the higher their alcohol level, the more diacetyl tones are present, effectively creating a closer link with the cheeses.

Try putting this to the test the next time you are serving blue cheeses. To accompany them, serve wine lovers' favorite gourmet beer, Samichlaus Bier, Schloss Eggenberg, Austria. Serve it in a fluted wine glass at a high temperature, around 14°C (about 57°F). This Christmas beer, brewed only once a year, contains 14% alcohol and boasts a powerfully aromatic nose and a voluminous mouth. It is full and spherical, ages quite well in the bottle, and after several years takes on a great complexity of roasted and spicy tonalities, with a caramelized touch reminiscent of a soy sauce reduction.

You may find this surprising, but your mouth will confirm it: fino and manzanilla sherries are both richly endowed with diacetyl and oxidative ketones, making them a natural fit for blue cheeses.

The fruity notes of the thiol family found in blue cheese are also the major aromatic signature of Sauvignon Blanc wines. So you should choose sweet wines such as Sauternes, which are also dominated by acetones—in particular, those with a high proportion of Sauvignon Blanc, such as the late harvest New World wines based on this grape variety.

Be aware that when it comes to the powerful Roquefort cheese, a pairing with Sauternes-type wines is not always conclusive. In fact, it rarely is!

As for the floral aromatic compounds of linalool, from the large terpene family, we must accord a special place to sweet, creamy wines based on Gewürztraminer and Muscat, which are richly endowed with terpenic tonalities.

In summary, with blue cheeses—whether Rassembleu, Bénédictin, or Ciel de Charlevoix from Quebec, the powerful Roquefort or the delectable Fourme d'Ambert from France, Italy's penetrating gorgonzola, or England's "king of cheeses," Stilton—you should serve a Sauternes or similar wine, such as a late harvest Sauvignon Blanc; a late harvest or SGN (Sélection de Grains Nobles) Gewürztraminer; an Austrian

Scheurebe (Gewürztraminer's molecular twin); a VDN Muscat wine; a fino sherry; a manzanilla sherry; or a New World oaked Chardonnay.

A vintage port should also have a privileged place beside blue cheese. If you're going with powerful blues, such as Roquefort or Stilton, it's best to serve a young vintage (about fifteen years old) that still has a fairly thick, velvety fruitiness, which will allow it to hold its own with the cheese.

Such is the case for the very harmonious, velvety, captivating Smith Woodhouse Vintage Port 1994, Symington Family Estates, Portugal, from what is considered one of the greatest vintages of the twentieth century.

On the other hand, if you go with a less powerful blue cheese, such as the delicious Fourme d'Ambert or the complicated Rassembleu, treat yourself to a seasoned vintage port, over twenty years old, such as the full and profound Dow's Vintage Port 1985, Symington Family Estates, Portugal, or the magnificent Graham's Vintage Port 1970, Symington Family Estates, Portugal.

Savored for the umpteenth time during the winter of 2009, the Graham's 1970 showed itself to be simply spectacular. It has the allure of a mature Burgundy grand cru, with the intense, peppery flavors of cherry pie and mint chocolate, a strong presence in the mouth, and the generosity of taffeta, all of which make it royalty in the kingdom of blue cheese.

**3.**

### THEME III
## COMPLEMENTARY
### FOODS
#### BLUE CHEESES

ANCHOVIES
COOKED ONIONS
COOKED SPINACH
DASHI
DRIED, AGED HAM (SPANISH HAM, PROSCIUTTO)
GALANGAL
GINGER

LYCHEE
MISO
MUSCAT GRAPES
PEARS
QUINCES
RECONSTITUTED DRIED MUSHROOMS (SHIITAKE, MOREL, OYSTER)

ROASTED SESAME SEEDS
SCALLOPS
SEAWEED (NORI, KOMBU)
SOY SAUCE
TURMERIC
YELLOW APPLES

# COMPLEMENTARY WINES AND BEVERAGES
**STRONG BLUE CHEESES** (SUCH AS ROQUEFORT AND STILTON)
**MODERATE BLUE CHEESES** (SUCH AS FOURME D'AMBERT AND RASSEMBLEU)

EXTRA-STRONG

15 YEARS OLD

VINTAGE PORT

MAURY

DARK BEER

VDN WINES

YOUNG BANYULS

STRONG BLUE CHEESES

RIVESALTES

FINO SHERRY

GEWÜRZTRAMINER

SCHEUREBE

LATE HARVEST

AUSTRIA

SWEET

SGN (SÉLECTION DE GRAINS NOBLES)

20 YEARS OLD AND OLDER

VINTAGE PORT

AND SIMILAR WINES

OAKED CHARDONNAY

SAUTERNES

EXTRA-STRONG

BELGIAN BEER

MODERATE BLUE CHEESES

ABBEY BROWN OR DARK

VDN WINES

SAUVIGNON BLANC

MANZANILLA SHERRY

LATE HARVEST

MUSCAT

# CINNAMON

ALMONDS
ANGELICA (ROOT)
BAY LEAVES
BERGAMOT
BITTER ORANGE
CARDAMOM
CHAMOMILE
CITRUS FRUIT
CLOVES

CUMIN
DILL
FRESH FIGS
GINGER
HOPS
HOT PEPPERS
LAVENDER
LEMON
LICORICE

PEPPER
ROSEMARY
SAFFRON
STAR ANISE
THYME
MINT
VANILLA
PEPPER
ROSEMARY
SAFFRON
STAR ANISE
THYME
VIETNAMESE CORIANDER
(LEAVES)
YUZU
VIETNAMESE CORIANDER
(LEAVES)

GRENACHE/
GARNACHA/
CANNONAU

OLOROSO
SHERRY

DILL
FRESH FIGS
GINGER
HOPS
HOT PEPPERS
LAVENDER
LEMON
LICORICE
MINT

RED WINES PRODUCED
BY CARBONIC MACERATION

GEWÜRZTRAMINER/
MUSCAT/PINOT GRIS

(ROOT)

# CINNAMON

## A HOT AND SENSUAL SPICE

"Science delivers its knowledge to those who seek it out."

HUBERT REEVES

Let's continue our exploration of the world of aromatic molecules, this time with a look at the volatile compounds responsible for the hot and sensual fragrances of cinnamon. By arranging the right combinations of cinnamon and its molecular allies in the domains of cuisine and wine, we have extraordinary new opportunities to delight the senses.

### THE MOLECULAR STRUCTURE OF CINNAMON

As I have already mentioned, the fragrance of a given spice or herb is not attributable to a single molecule; rather, it is composed of a cocktail of volatile molecules, in variable proportions, which confer an ultimate aromatic signature. It is also important to note that, as in the case of wine, the terroir, climate, and growing methods significantly influence the relative proportions of aromatic compounds in spices and herbs.

Occasionally, certain aromatic compounds dominate the others, in quantity or in power, and so define the ingredient's major tone; this is the case, for example, with cinnamic aldehyde (also called cinnamaldehyde) for cinnamon.

Cinnamon also contains ethyl cinnamate, an ester whose fruity balsamic fragrance helps give cinnamon its signature aroma. Ethyl cinnamate, as we have seen, also occurs in strawberries. *Eucalyptus olida* (strawberry gum) contains a very high percentage of a related ester, methyl cinnamate (see the "Ginger" chapter). Because of this, the leaf of *E. olida* is

sometimes used to reproduce the fragrance of cinnamon in the food and perfume industries.

Ethyl cinnamate also develops in red wines that undergo carbonic maceration, such as Beaujolais Nouveau.

Various cinnamon varieties contain sesquiterpenes, such as alpha-caryophyllene (which has a woody fragrance) and humulene (also called beta-humelene and alpha-caryophyllene, and found in hops and Vietnamese coriander).

We also find terpenic alcohols in cinnamon, such as gamma-terpineol (pine needles/bitter orange), borneol (camphor or woody fragrance, present in chartreuse, coriander, eucalyptus, bay leaves, rosemary, and savory), and linalool (lavender/lilies of the valley), as well as benzaldehyde (bitter almonds).

Cinnamon also contains, in smaller quantities, coumarin (vanillin aroma), eugenyl acetate (anise aroma), dihydrocapsaicin (one of the burning molecules in hot peppers), and safrole (found in sassafras, among other plants).

These various compounds all participate in forming cinnamon's aromatic signature, and we can use them as our building blocks for devising new wine and food pairing ideas. In general, cinnamon's flavor is sweet, becoming almost hot, sometimes even burning, vaguely bringing to mind cloves and pepper.

# 1.

## PRINCIPAL COMPLEMENTARY FOODS
### CINNAMON

| | | |
|---|---|---|
| ALMONDS | DILL | PEPPER |
| ANGELICA | FRESH FIGS | ROSEMARY |
| BAY LEAVES | GINGER | SAFFRON |
| BERGAMOT | HOPS | STAR ANISE |
| BITTER ORANGE | HOT PEPPERS | THYME |
| CARDAMOM | LAVENDER | VANILLA |
| CHAMOMILE | LEMON | VIETNAMESE CORIANDER |
| CITRUS FRUIT | LICORICE | |
| CLOVES | MINT | YUZU |
| CUMIN | PASTIS | |

# 2.

## SECONDARY COMPLEMENTARY FOODS
### CINNAMON

| | | |
|---|---|---|
| AMARETTO | KIRSCH | SCOTCH |
| APRICOTS | MALTED BARLEY | STRAWBERRIES |
| BEER | MANGOS | THAI BASIL |
| CHARTREUSE | MOZZARELLA | TOBACCO |
| CHERRIES | NUTS | TONKA BEANS |
| COOKED ASPARAGUS | PEACHES | WILD BASIL |
| EUCALYPTUS | PRUNES | |
| GRAPES | RASPBERRIES | |
| GRILLED BEEF | SAVORY | |

## CINNAMON'S TWO MAIN VARIETIES

Cinnamon comes from the bark of cinnamon tree branches. This bark naturally curls up as it dries, forming a hard, pale brown stick in the form of a double spiral. The cinnamon tree is widely cultivated around the world, but the best cinnamon comes from its native land, Sri Lanka. Although there are many varieties, we can basically group them into two large families, Chinese cinnamon and Sri Lanka (Ceylon) cinnamon.

### CHINESE CINNAMON *(CINNAMOMUM CASSIA)*

Chinese cinnamon, with the scientific name of *Cinnamomum cassia,* is the best known and most widely consumed variety. It is readily found on the market in the form of sticks.

Despite having a chemical composition similar to Ceylon cinnamon, it is characterized by the strong presence of cinnamic aldehyde (50% to 75%), as well as coumarin, but it contains no or very little eugenol, the major compound in cloves.

The seasoned amateur will find Chinese cinnamon cruder than the Ceylon variety, even though most consumers find it more interesting, more powerful, and hotter—I would say more physical and immediate, but also more one-dimensional. Chinese cinnamon is known better by its prickly taste than by its fragrance.

### CEYLON CINNAMON *(CINNAMOMUM ZEYLANICUM)*

Cinnamon from Sri Lanka (formerly Ceylon), also known as "true cinnamon," contains substantially fewer phenolic compounds (cinnamic aldehyde) than the Chinese variety, and also very little coumarin. Ceylon Cinnamon is subtler and more complex than its cousin. It contains, among others, notes of linalool (floral fragrance) and eugenol (cloves make up 4% to 10% of its molecular weight). Both linalool and eugenol are virtually absent from the Chinese variety.

## CINNAMON AT TABLES AROUND THE WORLD

Cinnamon has become one of North America's favorite spices, especially for apple and chocolate desserts and for flavored coffees.

In Europe, cinnamon is especially common in compotes and jams. During the Middle Ages, it was used to make hypocras, a wine sweetened with a generous amount of honey and spiced with ginger and cinnamon.

In the Middle East, it flavors meat dishes such as tajine, while in India it appears in some curries.

In Asia, cinnamon is often part of spice mixes, such as the five-spice powder popular in Chinese cuisine, which is most commonly made up of cinnamon, star anise, cloves, Sichuan pepper, fennel, and sometimes licorice root. In varying quantities, cinnamon also enhances dishes featuring duck, pork, and chicken, as well as Chinese curries and Singapore curried beef and noodles, accompanied with a spicy *sambal.* And, of course, it is often used for perfuming teas as well.

Chefs all around the world have been according cinnamon an increasingly important place in their savory creations, such as in lamb dishes. Cinnamon's presence in savory foods is usually subtler than it is in desserts.

## CINNAMON: A FLAVOR ENHANCER

It's a little-known culinary secret that cinnamon, when used sparingly, can be a flavor enhancer, amplifying other flavors, just as can sherry, coffee, capsaicin, asparagus, and glutamate. If you use cinnamon subtly in your recipes, it will add a real presence.

## CINNAMON IN WINE'S AROMATIC CONSTITUTION

In wines, a cinnamon aroma can come either from the oak of the barrels (for barrel-raised wines) or from the lignins in the grape stems, in cases where the stems were not removed or were removed just prior to fermentation.

The lignins in the stems of Grenache grapes, especially those from the Rhone Valley and Languedoc-Roussillon, are richly endowed with aromatic compounds that can produce alcohol and cinnamic aldehyde, a compound that plays a large role in the flavor and fragrance of cinnamon.

Both white and red wines raised in new barrels may contain these molecules that furnish a cinnamon aroma. This is particularly the case for Southern Hemisphere and Californian Pinot Noirs, and Spanish red wines from Ribera del Duero and Rioja.

Cinnamon and its aromatic derivatives can also be detected in red wines that have undergone carbonic maceration, such as Beaujolais Nouveau. This winemaking technique, which

3.

COMPLEMENTARY
WINES
**CINNAMON**

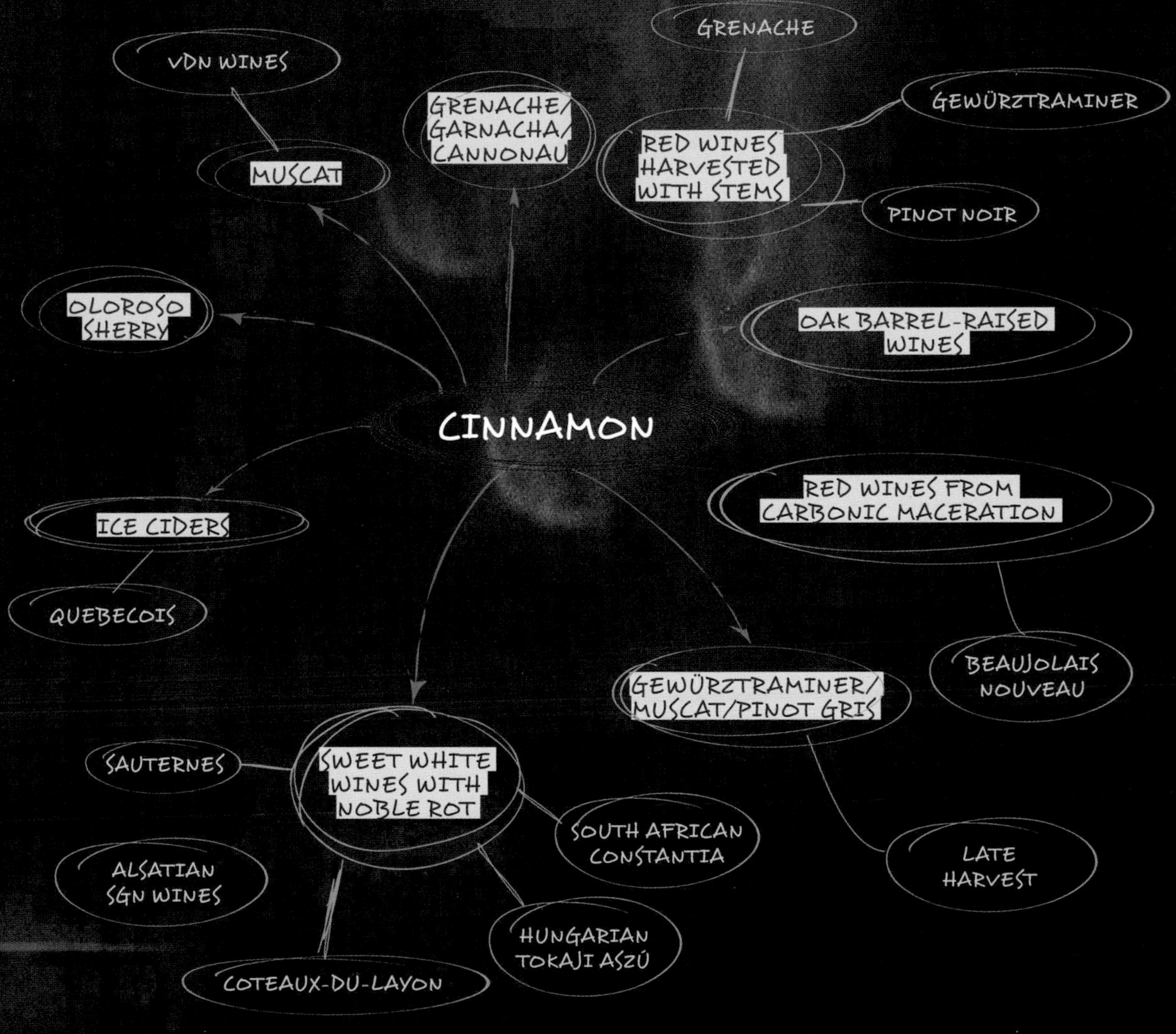

# CINNAMON

- GRENACHE
- GEWÜRZTRAMINER
- RED WINES HARVESTED WITH STEMS
- PINOT NOIR
- GRENACHE/GARNACHA/CANNONAU
- VDN WINES
- MUSCAT
- OLOROSO SHERRY
- OAK BARREL-RAISED WINES
- ICE CIDERS
- QUEBECOIS
- RED WINES FROM CARBONIC MACERATION
- BEAUJOLAIS NOUVEAU
- SWEET WHITE WINES WITH NOBLE ROT
- SAUTERNES
- ALSATIAN SGN WINES
- COTEAUX-DU-LAYON
- HUNGARIAN TOKAJI ASZÚ
- SOUTH AFRICAN CONSTANTIA
- GEWÜRZTRAMINER/MUSCAT/PINOT GRIS
- LATE HARVEST

involves processing whole grapes in a sealed vessel containing carbon dioxide but no oxygen, develops ethyl cinnamate in the wine. This ester, originating in the combination of cinnamic acid and ethanol, has a fruity, balsamic aroma, reminiscent of cinnamon with a touch of amber.

We also find these precious aromatic molecules in white Alsatian wines, especially late-harvest Gewürztraminers and, to a lesser extent, those based on Pinot Gris and Muscat. VDN Muscat wines, raised mostly near the Mediterranean Sea, are also influenced by such molecules. Other affected wines include sweet whites based on Sémillon (Sauternes) or Chenin Blanc grapes that have been touched by noble rot (botrytis cinerea). Examples include Coteaux-du-Layon from the Loire Valley, SGN (Sélection de Grains Nobles) wines from Alsace, Tokaji Aszú from Hungary, and South African Constantia.

And let's not forget Quebec ice cider, which is made by fermenting the juice of frozen apples. This sweet cider also shares compounds that are found in cinnamon's aromatic profile.

Finally, note that sherry contains cinnamic acid (cinnamon), benzaldehyde (almonds), and coumarin (vanillin), all three of which are major constituents of cinnamon. They appear in some fino sherry fragrances, but the longer that sherry matures in casks, the higher the level will be of these three volatile compounds. Thus, it's in oloroso sherry, and in some amontillados, that the aromatic aromas associated with cinnamon are the most noticeable.

## SOME WINE AND FOOD PAIRING POSSIBILITIES WITH CINNAMON

To magnify the sweet spice notes and make the most of the silky texture of a very enchanting southern Pinot Noir, such as the Coldstream Hills Pinot Noir 2007, Yarra Valley, Australia, with a body that is simultaneously full and vaporous, select a dish such as pork fillets with cinnamon and cranberries. The soft, spicy fragrances and the sweet-and-sour flavor of the honeyed cranberries call for a Pinot Noir from California or Australia.

Without leaving the charming kingdom of Pinot Noir, we also have the irresistible Kim Crawford Pinot Noir 2007, Marlborough, New Zealand. This wine, bursting with fruits and flavors, with tender tannins, is very fresh and has an enveloping body. Juicy, dazzling, and delicious, filled with long flavors of black cherries and sweet spices, it's a great sensation, especially with five-spice chicken roasted with sesame seeds or a five-spice Vietnamese pork sauté.

For spicier cinnamon-dominated dishes—such as a lamb tajine with five spices and caramelized cipollini onions, or beef short ribs in cinnamon and red wine curry (a recipe from Vij's Elegant and Inspired Indian Cuisine by Vikram Vij)—opt for a Rhone Valley or Spanish Grenache-based wine, such as Monasterio de Las Viñas Gran Reserva 2001, Grandes Vinos y Viñedos, Cariñena, Spain, made from 50% Garnacha, 30% Tempranillo, and 20% Mazuelo grapes. This is a quite compact, condensed wine, whose aroma starts off discreet and becomes more complex and detailed after oxygenating in a carafe, revealing scents of Havana tobacco, black cherries, cinnamon, nutmeg, and a very light touch of wood. It has a certain density in the mouth, presenting tannins wrapped in a creamy coating.

## SOMMELIER-COOK'S HINT

**Cinnamon-infused olive oil** Some olive oils adapt better than others to the infusion of spices. One example of an olive oil that does so beautifully is the Spanish Cornicabra, an oil that is soft and pleasant, yet profound at the same time. This olive oil is ideal for infusing with exotic spices, including ginger, cinnamon, cardamom, and pink pepper, or for adding to a fruit sorbet. Infused with cinnamon, it's perfect on desserts and pastries, as well as on game.

To prepare the spiced olive oil, place crushed cinnamon or cinnamon sticks in a small bottle. Pour the oil on the spices. Once the infusion is finished, remove the spices and set aside the infused oil. Dilute it with plain oil before each use. The infused oil can be used for about two weeks.

At cheese or dessert time, select a Sauternes-style wine, so a wine affected by noble rot (botrytis cinerea), which pairs perfectly with cinnamon. One fine example is the exuberant, spicy, preserved, and unctuous Château Bel-Air 2004, Sainte-Croix-du-Mont, France, which serves as a bridge between the cheese and the dessert. Start with a cheese plate accompanied by figs roasted in cinnamon and honey.

Finally, finish off your meal with fireworks, courtesy of a millefeuille of gingerbread and apples (recipe in *À table avec François Chartier*) accompanied by the same wine, or, even better, a bracing Quebec ice cider, whose bouquet has some molecular similarities with that of our hot and sensual cinnamon.

**SOMMELIER-COOK'S HINT**

**Figs roasted with cinnamon and honey** Stuff quartered figs with a paste of softened butter, honey, and cinnamon, then cook for 5 minutes in a preheated oven at 204°C (400°F), taking care to baste the figs regularly with the paste during cooking. Accompany this dish with either a sweet white wine affected by noble rot *(botrytis cinerea)*, or an oloroso sherry.

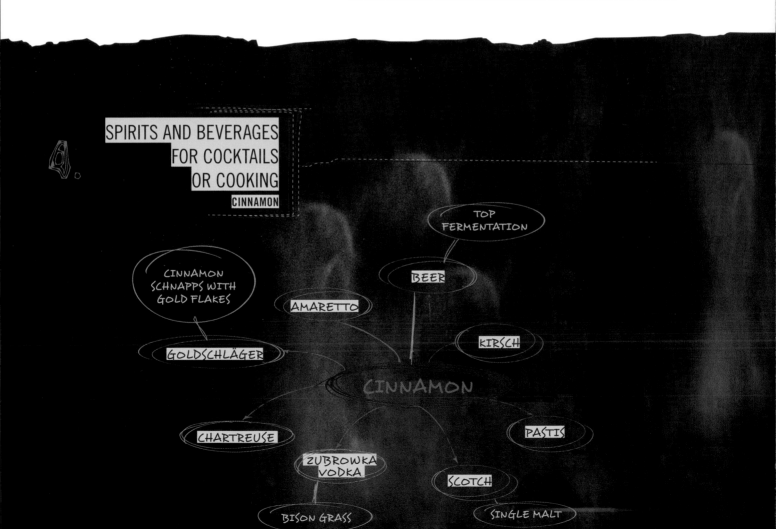

SPIRITS AND BEVERAGES FOR COCKTAILS OR COOKING

CINNAMON

CINNAMON

TOP FERMENTATION

BEER

CINNAMON SCHNAPPS WITH GOLD FLAKES

AMARETTO

KIRSCH

GOLDSCHLÄGER

PASTIS

CHARTREUSE

ZUBROWKA VODKA

SCOTCH

BISON GRASS

SINGLE MALT

SOLUBLE IN FATTY AND SUGARY SUBSTANCE AND IN ALCOHOL.

INSOLUBLE IN WATER AND ACIDIC BEVERAGES.

...ATELY, THE ...ON OF HEAT ...ROVOKES ONLY ...OUT FIFTEEN ...NUTES.

CAPSAICIN
(HOT PEPPERS)

CORIANDER

ROASTED ALMONDS

# CAPSAICIN

## THE "FIERY" MOLECULE IN CHILI PEPPERS

"The absence of proof is not a proof of absence."

HUBERT REEVES

From the subtle Espelette chili pepper to the extra-hot Bhut Jolokia, by way of Cayenne peppers, paprika, hird peppers, pimentón, chipotles, Serrano peppers, New Mexico chilis, habanero peppers, and the powerful jalapeño, there's a whole world of spicy sensations to discover!

There are several hundred varieties of chili peppers in the world, including sweet peppers—all members of the *Capsicum* genus in the Solanaceae family—and some three thousand hybrids. In Mexico alone, about 350 types of chilis have been classified.

Chili peppers are enjoyed around the world, from chili con carne in Mexico, to green curry in Thailand, to satay and *sambal* in Indonesia, and a range of cuisine in the Sichuan and Yunan provinces of China. Hungarians have their paprika and the Spanish their pimentón, both of which are fairly mild peppers.

North Americans, for their part, have the famous Tabasco, a sauce based on the very hot tabasco pepper, which is mixed with salt and vinegar and macerated for several years in oak barrels. Some varieties are quite aged, and others extra-strong. Tabasco is used to spice up cocktails and recipes, even in the kitchens of some of the world's greatest chefs.

The recent popularity of "Tex-Mex" cuisine has made chili the most widely consumed spice in the world, with a production that is twenty times greater than that of the famous black pepper!

### CAPSAICIN

Chili peppers of the *Capsicum annuum* family, the most widespread variety of hot peppers, are rich in various volatile compounds that give them their aromas and their flavors, including capsaicin and dihydrocapsaicin, as well as some carotenoids.

It's capsaicin and dihydrocapsaicin (the latter also occurs in cinnamon; see the chapter of that name), two alkaloids of the capsaicinoid family, that are the major "fiery" molecules of various chili pepper varieties. They give hot peppers their unique sensation of burning in the mouth. These compounds are mostly found in the peppers' pulp and seeds.

### A REAL KICK

Capsaicin is the food world's most irritating compound. On the Scoville scale (see page 184), in its pure state, capsaicin has a value of 16 million Scoville heat units (SHU), in contrast to ginger's shogaol (160,000 SHU), pepper's piperine (100,000 SHU), and ginger's gingerol (60,000 SHU).

Contrary to popular belief, capsaicin has no chemical effect on the taste buds, as do, for example, the tannins in red wine. In other words, it doesn't physically burn. Rather, it has a neurological effect on the brain, via the nerve endings (the trigeminal nerve), provoking the secretion of endorphins, hormones associated with well-being. This explains in part why humans take such pleasure in consuming hot peppers, even

1.

**IMPACT ON TASTE**

CAPSAICIN (HOT PEPPERS)

THE SENSATION OF HEAT THAT IT PROVOKES ONLY LASTS ABOUT FIFTEEN MINUTES.

INSOLUBLE IN WATER AND IN ACIDIC BEVERAGES.

SOLUBLE IN FATTY AND SUGARY SUBSTANCES, AS WELL AS IN ALCOHOL.

THE CARBON DIOXIDE IN CARBONATED DRINKS, BEER, AND SPARKLING WINE INCREASES THE "BURNING" SENSATION.

## CAPSAICIN
### (HOT PEPPERS)

ALCOHOL CALMS ITS HEAT (UP TO ABOUT 14%).

USED IN SMALL DOSES, IT BECOMES A FLAVOR ENHANCER.

IT IS ALSO FOUND IN SMALL QUANTITIES IN CINNAMON, CORIANDER, AND OREGANO.

IT IS A COUSIN OF GINGEROL, WHICH IS FOUND IN GINGER.

the fieriest ones—in contrast to wild animals, who flee from consuming them!

### A RELATIONSHIP WITH GINGER
Chemically, capsaicin is a close relative of gingerol, one of the principal spicy molecules in ginger (see the chapter of that name).

There are different capsaicin compounds, which explains the differences in taste among chili peppers. A pepper's richness in capsaicin, and thus its level of spiciness, depends in part on the genetic composition of each variety, but even more on the growing conditions and maturity of the pepper. This also explains the variations in spiciness that can occur even among hot peppers within the same family.

Similar to cinnamon, coriander, and oregano, which are also rich in capsaicin, although to a lesser extent, the capsaicin in chili peppers activates the heat receptors in the skin, thus provoking a pseudo-sensation of physical heat in the mouth. This can simulate a temperature exceeding 42°C (almost 108°F) and even cause burning, despite the fact that there is no true temperature increase. Fortunately, this strong burning sensation is only temporary, and greatly diminishes after about fifteen minutes.

### CAPSAICIN: THE OPPOSITE OF MENTHOL
Capsaicin has the opposite effect of the sensation created by foods that contain molecules such as anethole (found in star anise and fennel), estragole (tarragon), and menthol (mint), which activate taste receptors with cool temperatures between 8°C and 28°C (about 46°F to 82°F), thus stimulating a cold taste (see the chapter "A Taste of Cold").

### A SUBTLE WEAPON AGAINST OBESITY?
It has been proven that capsaicin provokes a feeling of satiety. It seems to be able to falsify the data sent to the brain so that when we eat a dish containing capsaicin, the brain gets the message that we are less hungry. This molecule also enables the body to burn more calories through the heat that envelops us while we eat it. Perhaps it can be used as an effective tool against the obesity epidemic…

### A DIET BUILT ON SPICY CUISINE?
Remember that capsaicin provokes the emission of endorphins (our natural "morphine") and thus provides us with a sense of pleasure. This explains the dependence that many cultures seem to have on very hot peppers, which almost seem to affect the senses like a drug. A diet based on spicy foods could thus unite the useful and the pleasurable!

### THE OTHER FLAVORS OF CHILI PEPPERS
Hot peppers also contain several other aromatic molecules that contribute to their unique flavor. The taste of peppers can be simultaneously fruity, sugary, and bitter, as well as peppery and aromatic, emitting volatile compounds from the same aromatic sphere as bananas, citrus fruits, citrus zest, sweet peppers, potatoes, and beets, among others. These latter three vegetables are marked by the vegetal aroma of methoxypyrazines (boxwood/green peppers)—just as are wines based on Sauvignon Blanc and Cabernet.

### DRIED AND SMOKED CHILI PEPPERS
Dried chili peppers have a greater aromatic complexity and are much more powerful than fresh chili peppers. Drying them concentrates their flavors and generates more varied fragrances. Some chili peppers are dried and smoked, such as Mexican chipotles and Spanish pimentón, generating unique flavors characterized by smoky tonalities that remain in the mouth for a long time.

It is worth mentioning that these smoky notes pair fantastically with wine, in particular with very minerally white wines and barrel-raised red wines. As we saw in the "Oak and Barrels" chapter, aging wines in barrels provides them with active molecules from the complex world of smoky flavors (burnt wood/burnt sugar/roasted tones).

### IS IT A FLAVOR ENHANCER?
According to popular belief, at least in Western cultures, the strong taste of chili peppers masks other flavors. But this isn't always the case. In fact, capsaicin in small doses stimulates certain nerves, slightly increasing our sensitivity to other tastes.

This explains the perfect union between hot peppers and dark chocolate. The Aztecs and the Incas were already onto this: they used hot spices, including chili peppers, to render their hot chocolate more complex.

This also explains why so many of us love to sprinkle our pizza and pasta with hot pepper–infused olive oil. The oil becomes a flavor enhancer, amplifying the flavors of the other ingredients, as if by magic.

A Bloody Caesar without Tabasco just wouldn't be as tasty. A single drop is enough to furnish presence and expressivity, proving once again the power of capsaicin as a flavor enhancer.

## MORE THAN HEAT!

Recent studies have demonstrated that chilis provoke in us more than the simple sensation of heat. Their sapid molecules create a temporary inflammation in our mouths, thus increasing the sensitivity of the taste buds and the mucous linings.

This increased sensitivity applies to temperature, touch, and the tactile or irritating aspects of some ingredients such as salt, acidic flavors, carbon dioxide, and cold.

This explains why once we are affected by the presence of capsaicin or piperine—the prickly molecule in pepper—our sense of taste becomes ultra-sensitive. This leads to the impression that the air we're inhaling is cooler than it actually is, and that the air we're exhaling is a lot warmer.

Wine served cold will seem even colder at such a moment; thus, it is important to adjust the serving temperature of wine in the presence of hot peppers.

On the other hand, too much capsaicin can reduce our sensitivity to the five basic tastes—salty, sweet, acid, bitter, and umami—as well as to fragrances. Remember, this is not really because of its physical heat, but rather as a result of redirecting the attention our brain usually pays to these flavors and aromas, which are easily recognized under other circumstances, when cooler heads prevail!

This underlines the importance of serving wines that are highly expressive, both in their aromas and in their taste, with dishes that feature hot peppers.

Finally, the more that one eats foods rich in capsaicin, the more desensitized one becomes to their effects. This is why Thailanders and Mexicans, for example, can tolerate certain dishes that the uninitiated would find inedible.

### TURNING ATTENTION AWAY FROM THE TASTE BUDS?

The deadly weapon that could completely counter capsaicin's fire does not yet exist. But we know that dry, solid, abrasive foods (rice, beans, soda crackers, or dry sugar, for example) temporarily distract the taste buds by sending new signals to the brain. This is why rice almost always accompanies Indian curries and Indonesian satays, and why red beans, with their slightly granular texture, so often accompany Mexican food.

### SPICY AND ACIDIC

 In order to maintain harmony in spicy dishes, one must add an acidic flavor, which will give élan and expressivity to the dish.

Certainly, hot peppers bring pleasure to those who love strong sensations, but beyond the sensation of fire, it's the subtle harmonic play of flavors that inspires some aficionados—in particular, Chinese from Sichuan, where spicy dishes mix with softer, more delicate ones.

Sichuan pepper is often used in Sichuan recipes as a catalyzing ingredient, calming the peppers' fire with its incomparable "electric flavor." Thanks to its organoleptic properties, it excites the taste buds into simulating a kind of electric sensation of crackling and stinging, reminiscent of the effect of putting your tongue on the post of a battery. The best varieties temporarily numb the mucous membranes and the lips, all the while perfuming the mouth with their lemony and floral tonalities, which recall the taste of dried rosebuds.

## 2. COMPLEMENTARY INGREDIENTS
### CAPSAICIN (HOT PEPPERS)

| | | |
|---|---|---|
| APRICOTS | CREAM | PINEAPPLE |
| ASPARAGUS | CURRY | POTATOES |
| BANANAS | DARK CHOCOLATE | RICE |
| BEETS | DRIED ROSEBUDS | ROASTED ALMONDS |
| BUTTER | FATTY SUBSTANCES | ROSEMARY |
| CARAMEL | FENUGREEK SEEDS | SOY SAUCE |
| CINNAMON | FRESH MANGO | STRAWBERRIES |
| CITRUS FRUIT | GRILLED BEEF | STRING BEANS |
| CITRUS ZESTS | MALTED BARLEY | SUGAR |
| CLOVES | MAPLE SYRUP | SWEET PEPPERS |
| COCOA | MILK | THAI BASIL |
| COCONUT (FRESH AND | MOZZARELLA | VANILLA |
| TOASTED) | OLIVE OIL | WILD BASIL |
| COCONUT MILK | OREGANO | YOGURT |
| CORIANDER | PEACHES | |

## THE ELECTRIC TASTE OF SICHUAN PEPPER

Sichuan pepper is perhaps the only ingredient known to provoke a taste that can be designated as electric (along with Sichuan buttons, flower buds that are actually native to Brazil but are named for their similarity in taste to Sichuan pepper). It is not a true pepper, but rather, rust-colored berries from the *Zanthoxylum* shrub that are dried and ground. Its flavor is close to pepper, especially when it is grilled just prior to use, and can also resemble anise. Sichuan pepper is also found in Chinese five-spice powder.

For dishes that feature hot peppers, red wine that's both generous and tannic, served cool, is a good choice. The astringency of the tannins in the red wine can serve the function of distracting the taste buds, as rice and beans do.

## ABOVE ALL, DON'T THROW WATER ON THE FIRE!

Contrary to common wine pairing wisdom, capsaicin from hot peppers is soluble in fatty matter and in alcohol, and not in water. This is why water and low-alcohol wines don't extinguish fire in the taste buds.

But fat, such as that contained in milk, butter, and ice cream, brilliantly succeeds in doing so. This explains why biologists studying the physiology of taste use milk to calm their taste buds when testing foods rich in capsaicin.

Capsaicin is even less soluble in dry, light, acidic white wines and in carbonated drinks.

Milk, bread, yogurt, cheese, butter, olive oil, ice cream, heavy sauces, and thick and unctuous wines—lightly acidic, ideally sweet, and containing a high level of alcohol (above 14.5%)—all succeed in calming capsaicin's intense heat. This is also true for ginger's shogaol and gingerol and black pepper's piperine when they are dominant in a dish.

Dishes containing capsaicin pair very well with certain sakes, such as Nigori, which is less rich in alcohol (10%) than other types of sake (at 17%, which is too high to calm the capsaicin). Nigori is lightly filtered, very rich in amino acids (and thus quite milky), and best served cold. With this type of sake, your Thai, Indonesian, Indian, and Mexican recipes will have never seemed so pleasant and refreshing!

## THE GUINNESS CHAMPION OF HOT PEPPERS

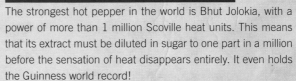

The strongest hot pepper in the world is Bhut Jolokia, with a power of more than 1 million Scoville heat units. This means that its extract must be diluted in sugar to one part in a million before the sensation of heat disappears entirely. It even holds the Guinness world record!

Sugar is more important than one might think where capsaicin is concerned. In 1912, to calculate the power of various hot peppers, an American pharmacist named Wilbur Scoville invented the Scoville scale. This scale is based on the dilution of chili peppers in a sugar syrup up to the point that the capsaicin no longer has a discernible effect on the tester's taste buds.

So, dare to serve sweet wines with spicy cuisine, Thai cuisine in particular, as well as dry, low-acid wines with a high alcohol content—alcohol that has a sweet taste. Sugar calms capsaicin much better than do acidity, cold, or especially the carbon dioxide in carbonated beverages—which actually increases capsaicin's impact. So if you like spicy foods, say adieu to beer and carbonated drinks…

As for the effect of alcohol which, past a certain level, has a sweet taste in the mouth, its heat joins up with that of the hot peppers and actually increases their impact. So once a wine's alcohol level exceeds approximately 14.5%, it needs to be served at a colder temperature. Or, even better, choose a wine whose alcohol doesn't exceed that limit.

## NO BEER!

Think twice before serving beer and carbonated beverages with cuisines spiced with hot peppers. Carbon dioxide intensifies and prolongs the burning effect of the capsaicin in the peppers, just as do black pepper's piperine and ginger's shogaol.

## VANILLA: A BUFFERING EFFECT

Vanilla has a buffering effect, as much in perfumery as in cuisine. It calms the intensity of other products that would be too volatile or too present in the mouth, whether due to bitterness or acidity. A very spicy dish may thus be softened by vanilla or by a wine with a strong vanilla taste.

Put another way, vanilla "rounds off" any acrid or spicy combination of foods. It is vanilla, in part, that softens the bitterness and fire of young cognacs and whiskies during their first years in the barrel.

For this reason, Spanish vintages from Rioja, Ribera del Duero, and Jumilla, raised either partially or completely in American oak barrels, are recommended. Other good choices are New World Merlots, Zinfandels, and Petite Sirahs.

## A QUESTION OF TEMPERATURE

As I mentioned previously, the temperature of a wine is also vitally important. A red that is both unctuous and generous must absolutely be served cold (about 14°C to 15°C, or about 57°F to 59°F) with dishes containing capsaicin. Because of the presence of this compound, the wine will seem almost icy.

It's the same for white wines. An unctuous, slightly acidic white wine with a moderate level of alcohol should be served at about 14°C (57°F) to allow its body to play the role of firefighter. But if the wine contains a lot of alcohol, lower the serving temperature to about 10°C (about 50°F), keeping in mind that it will slightly intensify the fire of the capsaicin, even if it seems almost iced. It's not easy to foil this molecule!

If your spicy dish is served very hot (temperature-wise), given that heat is a flavor enhancer for capsaicin, its high temperature will enhance your sensitivity to the fire. On the other hand, if the same dish is served cold or lukewarm, the burning sensation will seem less intense.

One must therefore take into account the dish's temperature, as well as that of the wine. The hotter the dish, the cooler the wine should be, without being iced. If the dish is served cold or at room temperature, you may serve the wine at a slightly higher temperature, but don't overdo it.

3.

## COMPLEMENTARY WINES TO CALM THE FIRE OF HOT PEPPERS
### CAPSAICIN (HOT PEPPERS)

**SEMI-SWEET/CREAMY WHITES**

GERMAN RIESLING
VOUVRAY
COTEAUX-DU-LAYON
SAUTERNES
JURANÇON
TOKAJI ASZÚ

**DRY WHITE WINES**

VIOGNIER

SÉMILLON BLANC
MARSANNE

**SAKE (SERVED COLD)**

NIGORI SAKE

**CAPSAICIN**

**ROSÉ WINES (SERVED WARMER THAN USUAL)**

BANDOL ROSÉ
TAVEL ROSÉ
VIN GRIS

**RED WINES (SERVED COOL)**

MONASTRELL
PETITE SIRAH
ZINFANDEL
GRENACHE/SYRAH/MOURVÈDRE

**RED WINES**

GARNACHA
TEMPRANILLO

RIPASSO
BACO NOIR
CARMENÈRE

CAPSAICIN
(HOT PEPPERS)

MENTHOL
EUGENOL
ESTRAGOLE
BORNEOL
ANETHOLE

SAUVIGNON BLANC (LOIRE/BORDEAUX/CHILE)
VERDEJO (RUEDA, SPAIN)
ALBARIÑO (RÍAS BAIXAS, SPAIN)
RIESLING (ALSACE/GERMANY)
ROMORANTIN (COUR-CHEVERNY, FRANCE)
CHENIN BLANC (SAVENNIÈRES/VOUVRAY, FRANCE)
COOL-CLIMATE CHARDONNAY (CHABLIS/NEW
ZEALAND)

# A TASTE OF COLD

## APPLES AND OTHER FOODS WITH REFRESHING FLAVORS

"Science is perhaps the only human activity in which errors are systematically criticized and, over time, corrected."

KARL POPPER

From apples to mint, by way of basil, tarragon, fresh fennel, green peppers, cucumbers, lemongrass, fresh coriander, parsley, lime, wasabi, daikon, eucalyptus, ginger, Chinese star anise, and raw celery, we now turn to foods with volatile compounds that provoke a cold taste in the mouth.

In this refreshing chapter, we'll follow the aromatic path of "cold-tasting" foods, starting with the apple, Quebec's signature fruit. It was the predominance of the apple's "cold" taste that inspired me to seek out other ingredients and wines that, due to certain shared volatile compounds, leave such a strikingly cool impression.

Even though the apple's origins, in central Asia, date back to more than sixty million years ago, and today there are between six thousand and ten thousand cultivated varieties, the apples we find in the marketplace today are unfortunately far from superior. In France, of the 20 kilograms (about 45 pounds) consumed each year by the average family, only 4 kilograms (about 9 pounds) are varieties other than Golden Delicious (about 40% of the total consumption), Granny Smiths, Braeburns, Galas, or Red Americans. The figures in Quebec are nearly the same.

What do we really know about the true pairing pleasures that cold-tasting foods and drinks, such as the apple, can offer in the kitchen, at the table, and in the glass?

### COLD IN THE WORLD OF TASTE

I chose the term "cold-tasting" to describe foods that contain volatile compounds, such as estragole in apples and menthol in mint (see the "Mint and Sauvignon Blanc" chapter), that provide them with a refreshing quality.

These molecules activate the taste receptors for temperatures between 8°C and 28°C (about 46°F to 82°F) and so simulate cold—the opposite effect, you may remember from the "Capsaicin" chapter, to that of the capsaicin in hot peppers, which simulates a temperature increase in the taste buds. This explains the sensation of coolness one feels in the mouth when eating an apple or mint, especially when they are raw.

Estragole, a cold-tasting molecule found in apples, also occurs in star anise, green basil, cinnamon, cloves, tarragon, ginger, fennel seeds, bay leaves, mustard, sage, and black licorice extract.

Several other foods from the anise-flavored family that includes apples and mint also possess this power to refresh. These include some of those mentioned above, including Chinese star anise, green basil, tarragon, and ginger, in addition to wild basil, carrots, celery and celeriac, lemongrass, cucumbers, fresh coriander, daikon, eucalyptus, fresh fennel, limes, lemon balm, parsnips, parsley and parsley root, green peppers, black radishes, horseradish, lemon verbena, and wasabi.

APPLES
BLACK RADISH
CARROTS
CELERIAC
CELERY
CHINESE STAR ANISE
CINNAMON

EUCALYPTUS
FRESH CORIANDER
FRESH FENNEL
GINGER
GREEN
GREEN PEPPERS
HORSERADISH

PARSLEY
PARSLEY ROOT
PARSNIPS
TARRAGON
WILD BASIL

All of these ingredients should be considered in dishes that feature apples. That is, of course, if you enjoy the cold taste!

### COLD-TASTING WINES?

Which wines should you serve with dishes dominated by apples, whether fresh or cooked, and other ingredients from the cold-tasting family?

Depending on a dish's other components, the presence of cold-tasting elements can have a calming effect. They can, for example, soothe the heat of spices, or seem to lower the temperature of a dish. So you can choose a wine rich in alcohol (alcohol gives the mucous membranes a sensation of heat) or serve it at a warmer temperature, because coolness is already accentuated in the mouth by the cold-tasting ingredients.

If your dish calls for a white wine, whether dry or sweet, raise the serving temperature slightly. In effect, the cold-tasting aromatic compounds increase the perception of cold, so that the wine seems colder than it is.

Cold-tasting foods reinforce the perception of acidity and bitterness in wine; here, the culprit is the temporary cold of the taste buds.

It is important to avoid serving white wines with biting acidity—although they are often suggested to accompany apples and ingredients from the apple family—and wines whose

## 1. COLD-TASTING VOLATILE COMPOUNDS
### TASTE OF COLD

MENTHOL
EUGENOL
ESTRAGOLE
BORNEOL
ANETHOLE

EUCALYPTOL
SINIGRIN
CAMPHENE
GERANIOL
LIMONENE

HEXANAL
R-CARVONE
APIGENIN

## 2. IMPACTS ON TASTE
### TASTE OF COLD

**3.** COMPLEMENTARY
INGREDIENTS TO
COLD-TASTING FOODS

TASTE OF COLD

APPLES
BLACK RADISH
CARROTS
CELERIAC
CELERY
CHINESE STAR ANISE
CINNAMON
CUCUMBERS
DAIKON

EUCALYPTUS
FRESH CORIANDER
FRESH FENNEL
GINGER
GREEN BASIL
GREEN PEPPERS
HORSERADISH
LEMON BALM
LEMONGRASS

LEMON VERBENA
LIME
MINT
PARSLEY
PARSLEY ROOT
PARSNIPS
WASABI
WILD BASIL

**4.** COLD-TASTING
WINES AND TEAS

TASTE OF COLD

SAUVIGNON BLANC
VERDEJO
ALBARIÑO
RIESLING
ROMORANTIN
CHENIN BLANC
COOL-CLIMATE CHARDONNAY

GRÜNER VELTLINER
RED VINHO VERDE
CABERNET FRANC
ICE CIDER
GYOKURO GREEN TEA

bitterness is too strong, unless you are one of the lucky few who adore bitter tastes.

Among dry white wines, choose wines dominated by Sauvignon Blanc, Riesling, Romorantin, or Grüner Veltliner, as well as Albariño or Verdejo. And let's not forget certain cool-climate Chardonnays raised in stainless steel vats, such as those from Chablis and from New Zealand.

For red wines, the choice is more limited. Cabernet Franc, in a fresh, airy version, is one of the few wines that possess both the cold taste and the structure necessary to stand up to the flavors of apple and its twin ingredients. Even though they are harder to find, some Portuguese reds carrying the Vinho Verde appellation are characterized by a striking cold taste.

Since the cold taste reduces the sweet sensation of syrupy white wines by increasing their acidity and bitterness, it effectively curbs the sugar in these wines, which unfortunately are all too rarely served at meals. Select sweet white wines based on the same grape varieties as those proposed for dry wines.

## JAPANESE GREEN GYOKURO TEA

The unusual cultivation method of Japanese green Gyokuro tea—three weeks prior to harvest, bamboo shades deprive the tea plants of sunlight—favors the important development of caffeine and carotenoids.

The result is a tea with a great complexity of volatile compounds from the carotenoid family, like those present in saffron and the wines that pair well with them. Also resulting from this process are volatile compounds in the sphere of cold-tasting foods, with tonalities of chervil. So serve green Gyokuro tea with dishes dominated by cold-tasting foods, as well as those based on saffron (see the chapter by that name).

## THE PHYSIOLOGY OF THE COLD TASTE

A dish entirely composed of cold-tasting ingredients, or dominated by one such ingredient, will reinforce the perception of acidity and bitterness in the wine that accompanies it. This is a basic principle of the physiology of taste.

In the presence of cold-tasting foods, a dry, nervy Riesling—one with a dominating acidity, and a more or less minerally taste—will seem to be biting and excessively acidic. A Spanish Verdejo from the Rueda appellation, with a fresh acidity, but generally more moderate, will increase in vivacity when paired with these foods; they also allow the wine's soft vegetal bitterness, usually relegated to the background, to take center stage.

As for white wines and cold-tasting ingredients, we want to choose whites that naturally cool the taste buds, but they need not be served very cold. Otherwise, the taste buds will be anesthetized by the cold taste provoked simultaneously by the foods and the wine, and by the wine's serving temperature.

A wine containing residual sugars, whether it be a semi-sweet, creamy, or dessert wine—for example, a Sauternes (a dessert wine)—will seem cooler and less sweet, while perhaps still leaving the impression of sweet bitterness as a finishing touch. Since its sugar is reined in by the cold taste, you may serve a Sauternes at any time during the meal, whether it's with an entrée or dessert.

## THE IMPACT OF THE "COLD" TASTE

The temporary anesthotizing effect of cold on the taste buds transforms the perception of wine's flavors; this is even more true if the wine is served very cold. The low temperature also slows down the wine's aromatic expression, as well as the unleashing and propagation of its flavors.

The cold taste brings acidic, bitter, and salty sensations to the forefront, and reinforces the tannins in red wines, at the same time diminishing the perception of sugars.

## SUMMARY: THE IMPACT OF THE COLD TASTE

+ A sweet wine served very cold seems less sweet than it really is.

+ A dry, lively white wine seems more acidic if served cold.

+ A tannic red wine served too cold seems more tannic than it really is.

+ A dish composed entirely of cold-tasting ingredients will bring forward the wine's acidity and bitterness (a dry Riesling, nervy and minerally, will become biting and explosive).

+ If a dish is served very cold, the dish should be light in acidity (lemon juice or wine) and in salty or bitter flavors.

+ If a wine is served too cold when paired with a dish, it will bring forward the food's acidic, salty, and bitter flavors, while calming the syrupy sweetness of sugars.

+ If a wine is served at a warmer temperature, the dish should be low in sugar; otherwise, the combination will be too heavy.

+ The higher the serving temperature, the more intense the perception of certain flavors will be; this is true for both food and wine.

## THE IMPACT OF HEAT

When there are no cold-tasting foods in a dish, or when the wine is served at a warmer temperature, we perceive the sweet tastes more and the bitter and acidic tastes less. Heat (which, like alcohol, can act as a solvent), in contrast to cold, acts as a flavor enhancer by more rapidly unleashing the aromatic molecules contained in herbs, spices, and wines.

## AT THE TABLE WITH COLD-TASTING FOODS

To put my theoretical research on cold-tasting foods to the test, cook yourself… a sandwich!

## SOMMELIER-COOK'S HINT

**A cold-tasting sandwich** Now that we know more about the cold taste and its relation to anise-flavored foods, here's a slightly different version of the sandwich from the "Mint and Sauvignon Blanc" chapter.

Make a goat cheese sandwich with thin, crunchy slices of green apples, green peppers, and cucumber, garnished with fresh mint and mayonnaise mixed with a touch of wasabi, and, if you so desire, a slice of smoked salmon. Serve with a fine glass of Sauvignon Blanc or Verdejo. Wine and food pairing has never seemed so easy and fulfilling!

From a pairing point of view, be assured that the ingredients in this sandwich work just as well with a red wine as with a white Sauvignon Blanc. Go for an expressive Spanish Verdejo, such as Verdejo Bribón Prado Rey, Rueda, Spain. This dry white wine expresses aromatic notes of fresh mint, fennel, and parsley, as well as of green apples and pink grapefruit. It is talkative, explosive, and crunchy in the mouth. Or go for a young, dashing, and flowing Chinon from the Loire Valley, such as Couly-Dutheil La Coulée Automnale, Chinon, France. This is an ideal example of the freshness and simple, direct radiance of Loire Valley Cabernet Franc. It has a festive taste and fine but very fresh tannins, with crisp acidity and flavors displaying radiance and presence.

Try making a striking cold soup, smooth and lightly creamy, from cucumber, potato, and lime zest, and serve it to your guests with a similarly inviting and zesty dry white Albariño wine, such as the minerally, compact, complex, and excellent Pazo de Señoráns, Rías-Baixas, Spain.

CHINESE STAR ANISE
CINNAMON
CUCUMBERS
DAIKON
EUCALYPTUS

STIMULATES A SENSATION
OF FRESHNESS IN THE MOUTH

## THE "TASTE OF COLD" OVER TIME

**IN THE NINETEENTH AND TWENTIETH CENTURIES:**

+ Sherberts
+ Ice creams
+ Cold soups
+ Cold jellies

**IN THE TWENTY-FIRST CENTURY:**

+ The new category of "cold-tasting" foods
  (Resulting from research in food harmony and molecular sommellerie)

+ Liquid nitrogen (freezing techniques and cold cooking)
  With the use of liquid nitrogen, foods can be frozen within a few seconds to about -180°C (-292°F), a kind of "polar frying." With nitrogen, the ice crystals are finer, releasing fragrances more rapidly into the mouth and offering a silkier texture than that of an ice cream that has been frozen in the freezer over the course of several hours. Don't forget that the intense cold of liquid nitrogen brings out the acidity and bitterness of several foods, such as avocados, that at first glance seem to contain no acidity or bitterness. Liquid nitrogen freezing requires sommeliers to revisit their standard wine pairings for selected foods. This technique, which is already being used in certain restaurants, is bound to make its way into our homes during the next fifteen years, much as the home fryer did in the middle of the twentieth century.

## YELLOW APPLES, CARROTS, SAFFRON...

As we saw in the "Saffron" chapter, yellow apples such as Golden Delicious owe their color to a high carotenoid content, in particular beta-carotene (this is also the case for saffron, carrots, yellow cauliflower, kombu and nori seaweed, and certain white wines). Red apples owe their color to anthocyanins, as do red grapes and, of course, red wine.

Possessing flavors associated with the cold taste, albeit more subtly than their green and red cousins, yellow apples transport us into another molecular universe.

For a truly harmonious pairing, try composing dishes using carotenoid-rich ingredients such as kombu and nori seaweed, carrots, yellow cauliflower, quinces, yellow apples, pears, and saffron, and pairing them with white wines that are just as rich in carotenoids (Chardonnay, Chenin Blanc, Sauvignon Blanc, and Riesling), or with cider (dry, still, sparkling, or even ice cider).

Here are two inspiring paths to try. First of all, a pork chop with Golden Delicious apples and saffron—a fine substitute for the traditional Christmas turkey—to be served with the penetrating Oyster Bay Chardonnay 2007, Marlborough, New Zealand. This inspiring dry white wine exudes exotic fruits in the aromatic sphere of pineapple and papaya. This wine remains exciting in the mouth, full and generous but also fresh and vibrant, with long-lasting flavors of strawberry and Golden Delicious apples, and a lightly floral scent reminiscent of saffron. You might also want to select a white Bordeaux, such as Château Haut Mouleyre Bordeaux 2005, Bernard Magrez, France, which has a surprising amplitude, creaminess, and generosity given its price. Its aroma, which requires substantial oxygenation in a carafe, is expressed by subtle yet profound touches of honey, saffron, and Golden Delicious apples. In the mouth, it has a unique texture and the persistence of a great wine, with a discreet acidity that gives way to a creamy thickness that is very rare in young white Bordeaux wine.

Finally, when it's time for dessert, prepare a delectable upside-down cake with yellow apples and saffron and fill your glasses with your favorite Quebec ice cider, such as those made by La Face Cachée de la Pomme, Cidrerie du Minot, or Cidrerie Michel Jodoin. It will beautifully echo this made-to-measure dish.

ANETHOLE

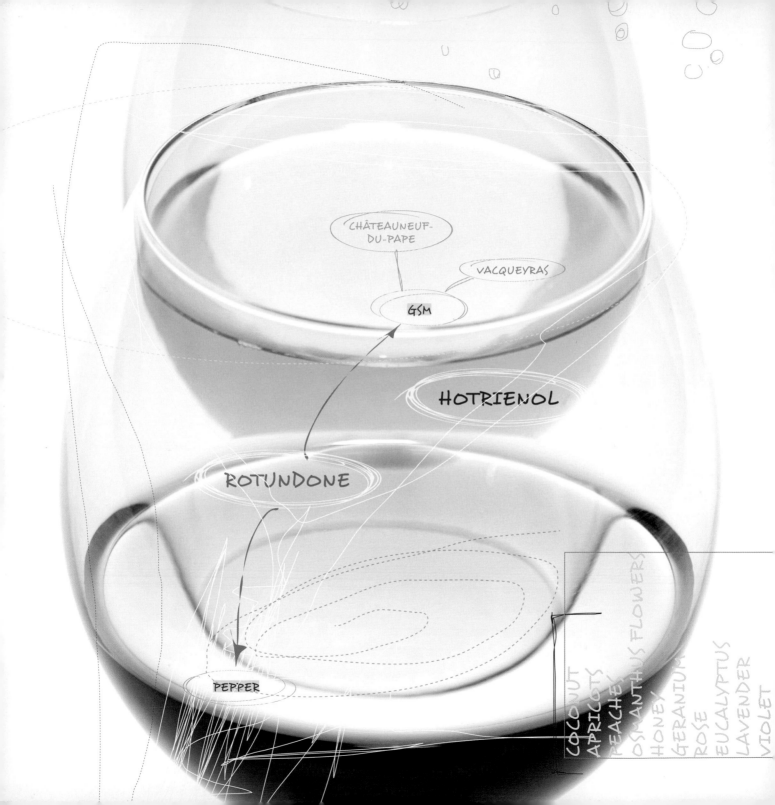

CHÂTEAUNEUF-DU-PAPE

VACQUEYRAS

GSM

HOTRIENOL

ROTUNDONE

PEPPER

COCONUT
APRICOTS
PEACHES
OSMANTHUS FLOWERS
HONEY
GERANIUM
ROSE
EUCALYPTUS
LAVENDER
VIOLET

# EXPERIMENTS IN FOOD HARMONY AND MOLECULAR SOMMELLERIE

## A MOLECULAR TASTING MEAL WITH TWO MASTER CHEFS

> "Methodology is necessary for researching the truth."
>
> RENÉ DESCARTES

To finish off, we look at three unique culinary events built around the research on wine and food harmonies that you have seen throughout this book.

In March 2009, with the cooperation of chef Stéphane Modat, formerly of the restaurant L'Utopie in Quebec City, I presented practical applications of my research on food harmony and molecular sommellerie. This was the grand premiere of my ideas, after more than three years of intensive research.

For the occasion, I invited the Bordeaux winemaker Pascal Chatonnet, with whom I have very close "molecular" links, as well as Thomas Perrin of the famous Château de Beaucastel, accompanied by his chef, Laurent Deconinck. These two wine-makers graced the table with a number of their wines.

### THE WORK OF PASCAL CHATONNET

This renowned winemaker, along with his wife, winemaker Dominique Labadie, is co-owner of the Excell laboratory in Bordeaux. Thanks to his research efforts, he was one of the important players in the identification and reduction of the infamous *Brettanomyces* yeast that can spoil wines by generating odors such as horse sweat. He was also the first person to explain clearly, in 1993, the origin of contamination by chloroanisoles and chlorophenols of wine cellars and, later, cork (TCA) and to find solutions for these problems. As if that weren't enough, he has earned two doctorates on the aromatic impact of oak barrels on wine.

Chatonnet is also the owner of several vineyards in Lalande de Pomerol, France, including Château Haut-Chaigneau and Château La Sergue in Libourne and Château L'Archange in Saint-Émilion.

A well-known "flying winemaker," Chatonnet brings his expertise to renowned wine estates all over the world, including Roda, Pintia, and Vega Sicilia (Spain), Stellenzicht (South Africa), J. & F. Lurton (Argentina), Tokaj-Oremus (Hungary), Sogrape (Portugal), Mandala Valley Vineyards (India), Mission Hill Family Estate (British Columbia, Canada), and in France, Château de Beaucastel (Châteauneuf-du-Pape), Château Montus and Château Bouscassé (Madiran), and Domaine Cauhapé (Jurançon).

OENOLOGIST PASCAL CHATONNET

Our summit meeting consisted of two parts, plus a third, more pleasurable activity, a sort of happy hour held at a wine bar. Let's start the account at the heart of our experiments in pairing harmony, with the gastronomic aspect, conceived around a magnificent meal created by two master chefs.

## A MOLECULAR TASTING MEAL WITH TWO MASTER CHEFS

For the occasion, we prepared a seven-course meal accompanied by nine specially selected wines. This was a "five-handed" meal, prepared by two master chefs, Laurent Deconinck and Stéphane Modat, inspired by my research, and in the presence of Thomas Perrin and Pascal Chatonnet and nine of their great wines.

For each course, with the help of Pascal Chatonnet, I identified one or several volatile molecules based on their aromas. I then set out to find complementary foods with a similar molecular signature—just as I did for the foods and wines we have explored throughout this book.

Stéphane and Laurent were thus able to count on ingredients that had a powerful attraction with each other and with the selected wines, creating a harmony of flavors both on the plate and between plate and glass. This was done for each course.

During the two months prior to these two tasting meals, we exchanged dozens of emails to inspire each other with ideas. Then the day before the events, we returned to the kitchen, where Pascal, the two chefs, and I confirmed that each course's ingredients were in harmony and polished up the wine and food pairings. It was an exploratory meeting that I would love to repeat as soon as possible!

## THE MEAL, STEP BY STEP

For the first course, built around a Coudoulet de Beaucastel Blanc 2006, Pascal Chatonnet singled out lactones (apricot and peach aromas) and hotrienol (linden and honey aromas). I suggested three ingredients complementary to these aromas: coconut milk, roses, and ginger. The wine and food pairing was thus accomplished with a "Bass poached in rose-flavored coconut milk, with marinated ginger with crunchy peas."

There are many other ingredients complementary to lactones and hotrienol that we could have worked with in the first course: roasted almonds, dried rosebuds, chamomile, snow crab, rosewater, fennel, prune kernel oil, jasmine, lavender, corn, honey, pastis, scallops, pork, aged Swiss cheese, aged Parmigiano-Reggiano cheese, and citrus zests.

The second course was inspired by the great white wine Château de Beaucastel 2006. We decided to create the wine and food pairing around the same two volatile compounds selected for the first course, as these aromatic molecules also occur in this white Châteauneuf-du-Pape, along with phenyl-acetaldehyde (honey aroma).

Because this wine was denser and more structured, we went with seafood rich in the umami flavor, which, because it creates presence and volume in the mouth, can support such a wine. The result: a penetrating "Large scallop warmed in an oil of bitter almonds, accompanied by a warm fennel salad with imperial mandarins and mirin, salted, dried corn powder, and osmanthus flower mousse."

The ingredients of this dish were all directly related to the three aromatic molecule families mentioned above.

### OSMANTHUS FLOWER: A DISCOVERY!

The osmanthus flower, here used dried, is used in China to add flavor to teas such as jasmine tea. It expresses pure tones of apricot and peach, influenced by the lactones present in its volatile structure. My research on lactone-rich foods led me to the discovery of this fantastic flower. With its great aromatic and harmonic powers, it deserves an honored place at the table.

For the third course, it was time for rotundone (pepper) and beta-ionone (violets) to enter the scene, having been chosen to accompany the red Perrin et Fils Les Christins Vacqueyras 2006 and, to a lesser extent, the Perrin et Fils Réserve Côtes-du-Rhône 2007 that accompanied it.

We had many complementary ingredients to work with that were derived from these two aromatic molecules, in order to enhance the cuisine and to showcase the two wines, especially the Vacqueyras: long-braised meat and ingredients rich in umami, kombu and/or nori, dried bonito, carrots, shiitake and matsutake mushrooms, juniper berries, dried herbs from southern France, black pepper, saffron, sage, red tuna, and aged, dried ham.

The result was a flavorful and unique third course: "Red tuna rubbed with juniper berries, black olives, beans, toasted nori confetti, melted cubes of ham fat, and a pipette of grape-seed oil and pistils of Moroccan saffron."

Nori contains beta-ionone (with a fragrance of violets),

which is soluble only in fatty substances; we included the oil-filled pipette with the third course so that guests could enjoy emptying it onto the nori, thus releasing the beta-ionone's precious violet flavor.

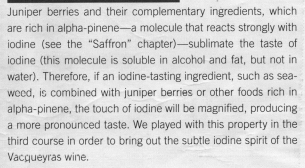

### A NOTE ON JUNIPER AND IODINE

Juniper berries and their complementary ingredients, which are rich in alpha-pinene—a molecule that reacts strongly with iodine (see the "Saffron" chapter)—sublimate the taste of iodine (this molecule is soluble in alcohol and fat, but not in water). Therefore, if an iodine-tasting ingredient, such as sea-weed, is combined with juniper berries or other foods rich in alpha-pinene, the touch of iodine will be magnified, producing a more pronounced taste. We played with this property in the third course in order to bring out the subtle iodine spirit of the Vacqueyras wine.

For the fourth course of our tasting menu, we "moved" from the Rhone Valley to Bordeaux in order to try out some of Pascal Chatonnet's wines. We served, side by side, a Château Haut-Chaigneau 2003 and a Château La Sergue 2006, two vintages from Lalande de Pomerol that Mr. Chatonnet suggested for us. The first wine contained beta-ionone (violets), frambinone (raspberries), and Havana extract (brown tobacco); the second contained maltol (torrified and woody aromas), guaiacol (woody and smoky aromas), and mercaptohexanol (black currant buds).

Thus was born a multi-faceted dish, made by joining two creations. On the left side of the plate, to accompany the Haut-Chaigneau 2003, we had: "On a jelly of raspberries glazed with violet oil, red sweet pepper morsels cooked in sesame seed oil, accompanied by kombu with licorice and hand-harvested sea salt smoked with Havana tobacco leaves and soaked in 1914 cognac." On the right side of the plate, to accompany the La Sergue 2006, we had: "Beef cold-smoked over maple wood, porcini mushrooms in a coconut and chicory butter, red cabbage crushed with black currant with boreal nutmeg, and stewing juice."

To the list of complementary ingredients sharing aromatic molecules with the two Lalande de Pomerol wines—which, again, were beta-ionone, frambinone, Havana extract, maltol,

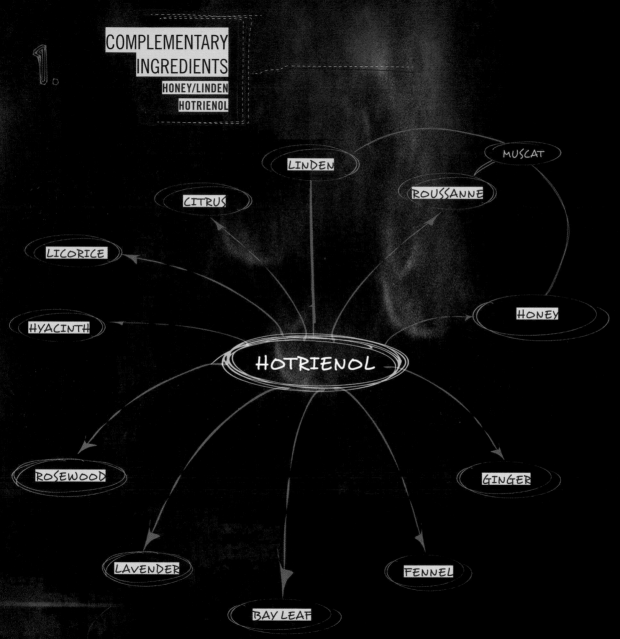

1.

**COMPLEMENTARY INGREDIENTS**

HONEY/LINDEN

HOTRIENOL

LINDEN

MUSCAT

CITRUS

ROUSSANNE

LICORICE

HONEY

HYACINTH

HOTRIENOL

ROSEWOOD

GINGER

LAVENDER

FENNEL

BAY LEAF

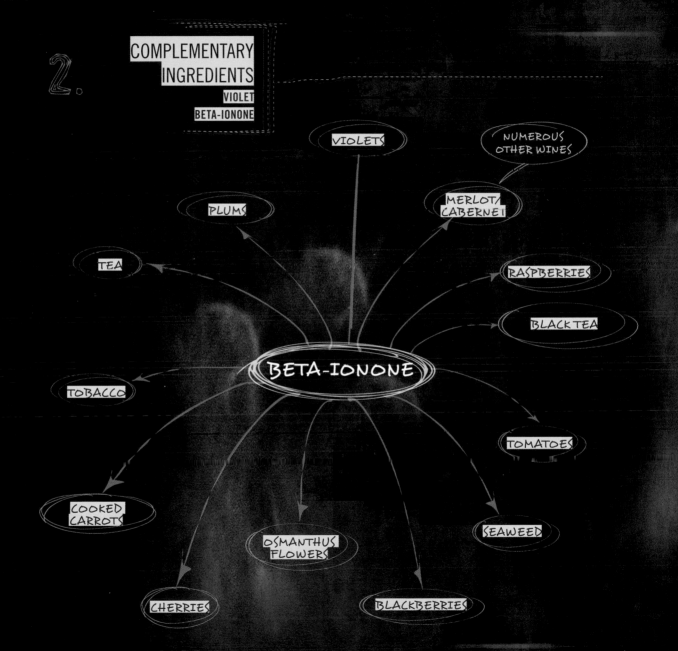

2.

COMPLEMENTARY
INGREDIENTS
VIOLET
BETA-IONONE

VIOLETS

NUMEROUS
OTHER WINES

PLUMS

MERLOT/
CABERNET

TEA

RASPBERRIES

BLACK TEA

BETA-IONONE

TOBACCO

TOMATOES

COOKED
CARROTS

SEAWEED

OSMANTHUS
FLOWERS

CHERRIES

BLACKBERRIES

3.

COMPLEMENTARY
INGREDIENTS
PEPPER (ROTUNDONE)

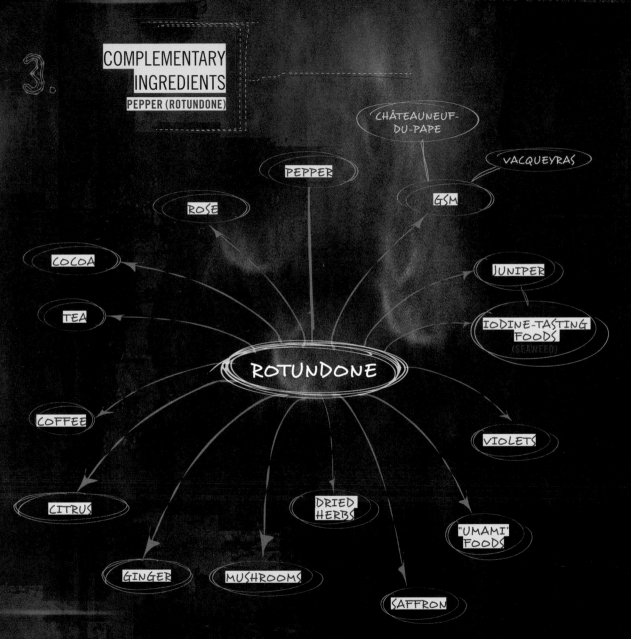

CHÂTEAUNEUF-DU-PAPE

PEPPER

VACQUEYRAS

ROSE

GSM

COCOA

JUNIPER

TEA

IODINE-TASTING FOODS
(SEAWEED)

ROTUNDONE

COFFEE

VIOLETS

CITRUS

DRIED HERBS

"UMAMI" FOODS

GINGER

MUSHROOMS

SAFFRON

TASTE BUDS AND MOLECULES

guaiacol, and mercaptohexanol—we had: seaweed, aromas from Havana tobacco leaves, balsamic vinegar (aroma only), smoked beef strongly grilled on the outside, 100% cocoa, coffee, raspberries, violet liqueur, fennel flower, toasted coconut, green peppers, plums, grilled sesame seeds, burnt sugar (unsweetened, such as maltol), and vandouvan.

The mature, intoxicating Château L'Archange 2001, a magnificent Saint-Émilion vintage by Pascal Chatonnet, defined the fifth course. Here, we worked with many of the same molecules that inspired the previous course, led by alpha-ionone (blackberry fragrance).

Our brainstorming, based on the results of my molecular research, led us to a "Magret duck breast cooked in smoked fat with wulong tea, marinated in wild carrot flowers and rose petals, with puffed wild rice and blueberries mashed with violets."

Once again, we only used foods that had a direct link to the volatile compounds of the L'Archange and the two previous Lalande-de-Pomerol wines, which share part of their aromatic profile with this Saint-Émilion.

### A NOTE ON CARROTS

During cooking, carrots generate new sapid molecules, including beta-ionone (with a fragrance of violets). This explains the use of carrots in this course.

The sixth course was inspired by a red Châteauneuf-du-Pape, a Château de Beaucastel 2005, which led us down the aromatic trail of spicy eugenol (the molecular signature of cloves) and dimethyl pyrazine (cocoa fragrance).

Around this great wine, we created our *plat de résistance*: "Inuit caribou, cooked in its juices with blackberry seeds, two types of celeriac purée (licorice and clove), honey fungus with cacao nibs, and Thai basil leaf."

The ingredients complementary to cloves and cocoa were: Thai basil, red beets, coffee, Ceylon (Sri Lanka) cinnamon, game (in particular, caribou), roasted sesame seeds, blackberry liqueur, hazelnuts, toasted coconut, barley, black pepper, potatoes, licorice, root beer, Scotch, asparagus roasted in olive oil and black pepper, and amontillado and oloroso sherries.

### A NOTE ON LICORICE EXTRACT CONCENTRATE AND YOUNG RED WINES

One of the great food pairings that I have discovered for wines based on GSM (Grenache/Syrah/Mourvèdre) blends, as was the Beaucastel, was game in a blackberry and licorice sauce. The black licorice extract concentrate (see the comments on licorice in the "Mint and Sauvignon Blanc" chapter), because of its unique structure, prolongs a wine and softens its tannins, in particular for young vintages such as our Beaucastel 2005. This is why we included black licorice extract concentrate in one of the celeriac purees that accompanied the caribou and the Beaucastel.

To conclude with a crescendo for the seventh course, I invited the two master chefs to recreate the profile of a Muscat wine in the form of a dessert. More precisely, we used the Perrin et Fils Muscat de Beaumes-de-Venise 2006.

So, departing from the volatile compounds dominant in the Muscat—namely, geraniol (geranium/rose/eucalyptus fragrance), linalool (lavender/citrus fragrance), and hotrienol (honey fragrance)—we created a delirious "Mascarpone Bavarois sweetened with orange honey in three aromatic versions: Geranium/Lavender, Lemongrass/Mint, and Eucalyptus."

This was a playful dessert, with different strata of juxtaposed flavors, all complementary with the wine because of their shared aromatic molecules.

The harmonious meeting of Muscat and lavender, designed by Stéphane Modat, will be remembered for a long time by the guests who were lucky enough to share in this excellent meal.

That covers the "how" of our experiment in molecular harmony, which brought together two master chefs and millions of taste buds…

### MOLECULAR TASTING: THE SCIENTIFIC SIDE

The molecular tasting (see menu on page 207), presented the day before the two-chef tasting menu, was offered exclusively to gastronomy and wine professionals. It was held at L'Utopie prior to the International Exposition of Quebec Wine and Spirits in the form of a scientific workshop on the volatile molecules in wine and food.

4.

COCONUT
APRICOTS
PEACHES
OSMANTHUS FLOWERS
HONEY
GERANIUM
ROSES
EUCALYPTUS
LAVENDER
VIOLETS
CITRUS FRUIT
STRAWBERRIES

RASPBERRIES
EXOTIC FRUITS
BLACKBERRIES
BLUEBERRIES
RAISINS
YUZU
LINDEN
LEMONGRASS
CARDAMOM
ASPARAGUS

CHICKEN
PORK
CLOVES
BUCKWHEAT
CHOCOLATE
CINNAMON
TEA
FINO SHERRY
COGNAC
RUM
BEER

MOLECULAR TASTING
CONCEIVED BY FRANÇOIS CHARTIER, WINE SCIENTIST
PASCAL CHATONNET, AND LAURENT DECONINCK
AND STÉPHANE MODAT, MASTER CHEFS

## 1. BEAUCASTEL BLANC 2006, CHÂTEAUNEUF-DU-PAPE, PERRIN ET FILS
### HONEY (PHENYLACETALDEHYDE) AND APRICOT/PEACH (LACTONES)

Large scallop warmed in an oil of bitter almonds, accompanied by a warm fennel salad with imperial mandarins and mirin, salted, dried corn powder, and osmanthus flower mousse

## 2. LES CHRISTINS 2006 VACQUEYRAS, PERRIN ET FILS
### BLACK PEPPER (ROTUNDONE) AND VIOLETS (BETA-IONONE)

Red tuna rubbed with juniper berries, black olives, beans, toasted nori confetti, melted cubes of ham fat, and a pipette of grapeseed oil and pistils of Moroccan saffron

## 3. BEAUCASTEL ROUGE 2005, CHÂTEAUNEUF-DU-PAPE, PERRIN ET FILS
### CLOVES (EUGENOL) AND COCOA (DIMETHYL PYRAZINE)

Fillet of venison marinated after cooking in red wine and a balsamic vinegar reduction, celeriac purée with cloves, honey fungus with cacao nibs, Thai basil leaves, and blackberry seeds

## 4. CHÂTEAU HAUT-CHAIGNEAU 2003, LALANDE DE POMEROL, PASCAL CHATONNET
### VIOLET (BETA-IONONE), TORRIFIED/WOODY (MALTOL), AND BROWN TOBACCO (EXTRACT OF HAVANA)

On a jelly of raspberries glazed with violet oil, red sweet pepper morsels cooked in sesame seed oil, accompanied by kombu with licorice and hand-harvested sea salt smoked with Havana tobacco leaves and soaked in 1914 cognac

## 5. CHÂTEAU LA SERGUE 2006, LALANDE DE POMEROL, PASCAL CHATONNET
### BLACK CURRANT BUDS (MERCAPTOHEXANOL) AND SMOKY/WOODY (GUAIACOL)
### SAME MOLECULES AS THE HAUT-CHAIGNEAU

Beef cold-smoked over maple wood, porcini mushrooms in a coconut and chicory butter, red cabbage crushed with black currant with boreal nutmeg, and stewing juice

## 6. CHÂTEAU L'ARCHANGE 2001, SAINT-ÉMILION, PASCAL CHATONNET
### TORRIFIED/WOODY (MALTOL), BLACKBERRIES (ALPHA-IONONE), VIOLET (BETA-IONONE)
### SAME MOLECULES AS THE HAUT-CHAIGNEAU AND LA SERGUE

Magret duck breast cooked in smoked fat with wulong tea, marinated in wild carrot flowers and rose petals, with puffed wild rice and blueberries mashed with violets

The tasting centered around the olfactory examination of aromatic compounds, in pure solutions, presented by the wine scientist Pascal Chatonnet in a comparative tasting between his wines and those of Château de Beaucastel.

For a grand finale, there were pairings between the wines and mini hors d'oeuvres, which had been conceived during the meetings between Laurent Deconinck and Stéphane Modat based on my research work on food harmony and molecular sommellerie. Six wines and six tapas, taken from the molecular tasting menu that I have just described, were served with a few small variations.

Pascal Chatonnet and I, supported by graphics illustrating my research, were able to put forth the theory of food harmony and molecular sommellerie over the course of these two activities. We exchanged ideas with the cooks, chefs, sommeliers, and reporters present, and by so doing, I hope, made advances in this new discipline.

## LARGE SCALLOP WARMED IN AN OIL OF BITTER ALMONDS, ACCOMPANIED BY A WARM FENNEL SALAD WITH IMPERIAL MANDARINS AND MIRIN, SALTED, DRIED CORN POWDER, AND OSMANTHUS FLOWER MOUSSE

A recipe by chef Stéphane Modat, formerly of L'Utopie in Quebec City, inspired by my research on food harmony and molecular sommellerie and presented at "A Molecular Tasting Meal with Two Master Chefs" in March 2009 (see the associated menu).

### INGREDIENTS FOR 8 PEOPLE

+ 8 large, fresh scallops (U/10)
+ One very firm fennel bulb
+ 5 cl mirin (about 3½ tbsp)
+ 2 imperial mandarin oranges (in season; otherwise, 2 tangerines)
+ 25 g dried corn
+ 20 g dried osmanthus flowers
+ 75 cl olive oil (about 3 cups)
+ Fleur de sel
+ Soy lecithin powder
+ Bitter almond extract

### PREPARATION
### VINAIGRETTE OF IMPERIAL MANDARIN ORANGES AND MIRIN

Blanch the mandarins in water three times, starting cold, and purée them in a mixing bowl. Add the mirin and the same quantity of cold water. Mix with 10 cl. of olive oil. Put through a fine Chinese sieve and place in the refrigerator.

### FENNEL PRESERVED IN OLIVE OIL

Throw away the first few fennel layers, which are somewhat tough. Using a mandoline, prepare a fine julienne by cutting the layers against the grain. Place them in a salad bowl, sprinkle with a good pinch of fleur de sel to season them well, and let stand for 10 minutes. Then rinse with water and place on a paper towel. Put a half liter (about 2 cups) of olive oil in a bowl and bring to 70°C (about 160°F), extinguish the heat, and add the fennel. Cover and let the mixture sit for half an hour, until the oil is lukewarm.

### OSMANTHUS FLOWER MOUSSE

Heat 1 liter (about 4 cups) of water in a bowl. Remove from the heat, add the dried flowers, and let the mixture steep for 5 minutes. Strain and add a scant teaspoon of soy lecithin. Mix with a hand-held mixer to obtain an airy mousse and place in a tall container.

### SCALLOPS

Make sure that the small tough nerve is removed. Cut the scallops in three, in the direction of their thickness. Place them on waxed paper, letting them overlap. Prepare the oil of bitter almonds by placing an espresso spoon of bitter almond extract in 25 cl. of olive oil. Coat the scallops with this oil before cooking.

### ASSEMBLY

Preheat the oven to 450°F (230°C). In a deep dish, place a full tablespoon of the lukewarm fennel mixture seasoned with a teaspoon of the imperial mandarin vinaigrette. Place the scallops in the oven on the middle rack and heat for only 3 minutes. Upon removing them from the oven, sprinkle with the dried corn powder, which you will have crushed in advance. Place the scallops on the fennel. Coat with the osmanthus flower mousse and serve.

## EXPERIMENTS IN FOOD HARMONY AND MOLECULAR SOMMELLERIE WITH CHEF STÉPHANE MODAT

Finally, on the occasion of the media launch of *La Sélection Chartier 2009*—the thirteenth edition of my annual wine guide—in October 2008 in Montreal and Quebec City, we set up two taste adventures centered on my research by presenting a "grand cuisine in miniature" menu, in the form of canapés served while the guests were seated.

I conceived and realized the menu with the friendly collaboration of Stéphane Modat, who came to Montreal along with his team. Following is the list of canapés and wine pairings that we presented, which I hope will inspire you to envision and create new wine and food pairings, starting with the same ingredients but cooked in your way.

## RED PEPPER PASTE WITH TOASTED SESAME SEED OIL AND GRILLED RED TUNA

This is another recipe created by Stéphane Modat on the basis of my work on food harmony and molecular sommellerie, and presented during the launch of *La Sélection Chartier 2009*.

### INGREDIENTS FOR 12 PORTIONS OF RED PEPPER PASTE
(According to the size of the cubes you make)

+ 500 g fresh red peppers
+ 25 g apple pectin
+ 25 g sugar
+ 1 tbsp toasted sesame seed oil
+ Red tuna (50 g per person)

### PREPARATION

Wash and cut the peppers in half lengthwise. Remove the stems and the seeds. Put the peppers in an oven dish and cover with parchment paper. Preheat the oven to 400°F (200°C). Cook the peppers in a hot oven for about 10 minutes. Then place them in a deep plate and cover with plastic wrap. Once the peppers are lukewarm, remove the skins, place them in a mixing bowl, and beat them to a smooth purée. Mix the sugar and the pectin and then mix this combination with the peppers before placing in a casserole. Cook at a low heat for about 5 minutes, stirring constantly.

When the cooking is finished, add the sesame oil, mix, place in a bowl, cover with plastic wrap, and refrigerate. Cut a good slice of tuna, salt it and lightly cover it with canola oil, and sear it on the grill just to mark it (it doesn't have to fully cook). Cut cubes from the tuna and from the red pepper paste and serve side by side.

FRANÇOIS CHARTIER
IN THE KITCHEN OF L'UTOPIE
WITH CHEF STÉPHANE MODAT

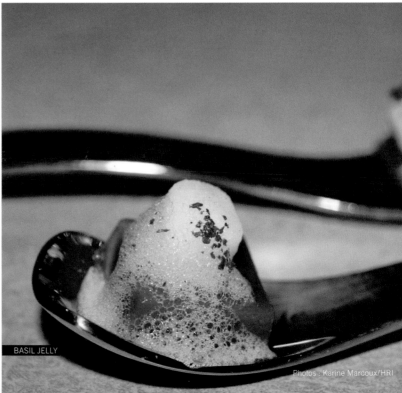

BASIL JELLY

Photos : Karine Marcoux/HRI

TASTE BUDS AND MOLECULES

MENU FOR THE LAUNCH OF *LA SÉLECTION CHARTIER*
CONCEIVED BY FRANÇOIS CHARTIER
AND STÉPHANE MODAT

**ANISE-FLAVORED/"COLD-TASTING"**

Cucumber lemongrass mousse accompanied by yellow carrot caramel and cubes of apple confit with zathar, served on a basil jelly

"A HINT OF VEGETABLE TASTE"
KABUSECHA KAWASE GREEN TEA
AND SPHERIFICATION OF JAPANESE CHERVIL

**ANISE-FLAVORED/"COLD-TASTING"**

Curly parsley meringue, fennel preserved with Chinese star anise, La Barre à Boulard creamy goat cheese (Tourilli Farm, Saint-Raymond de Portneuf, Quebec), organic yuzu juice, and fleur de sel with kaffir lime

**ANISE-FLAVORED/"COLD TASTING"/PYRAZINE**

Minted green pea marshmallow, vacuum-cooked pheasant in cedar, lovage mousse, and fleur de sel with kaffir lime

VERDEJO BRIBON 2006
RUEDA, PRADO REY, SPAIN

**VEGETAL PYRAZINE/LIGNANS**

Red pepper paste in sesame oil, roasted red tuna marinated in niora chili and sweet rice vinegar and served cold

**COOKING PYRAZINE/LIGNANS/VANILLIN/EUGENOL**

Beef smoked prior to cooking, puffed wild rice with chicory, Thai basil leaves, grapeseed oil perfumed with Jamaican peppers

CHÂTEAU GARRAUD 2005
LALANDE DE POMEROL,
VIGNOBLES LÉON NONY, FRANCE

**VEGETAL PYRAZINE/COOKING PYRAZINE/LIGNANS/ VANILLIN/EUGENOL**

Red beets preserved in vanilla, orange sections with Campari and cloves, fleur de sel with Trinidad cocoa, sumac powder

HENRIQUES & HENRIQUES
SINGLE HARVEST 1995
MADEIRA, PORTUGAL

**SOLERONE/VANILLIN/COUMARIN/EUGENOL/ LACTONES/BENZALDEHYDE**

Date cookies soaked in sherry and cinnamon, with a canopy of torrified coconut milk, butterscotch, Scotch, and almond milk

**SOLERONE /LACTONES/PYRAZINE/FURFURAL/ COUMARIN**

Fig squares, smoky cream, brown sugar with licorice root

"SMOKY FLAVORS"
WULONG TEA, PINGLIN BAO ZHONG 1985, TAIWAN

PYRAZINES

# MINI GLOSSARY

Our intention here is not to provide a full glossary of wine terms, but to define a few wine and food terms used throughout this book, particularly French terms, as well as to define the study of molecular sommellerie.

**Lees**—The sediment (consisting mainly of grape particles and dead yeast) that accumulates in the bottom of a wine container during fermentation. Some wines are aged "on their lees" (*sur lie*) to add complexity, body, and flavor.

**Molecular sommellerie**—The practice of developing food pairings and food and wine pairings based on dominant aromatic compounds; for example, a goat cheese sandwich with sliced apples, cucumbers, fresh mint, and Sauvignon Blanc wine. As this example indicates, such pairings may involve relatively simple foods.

**Siphon**—A device that adds air to a food, creating a foam. Whipped cream siphons and soda siphons have been used by molecular gastronomists to create interesting food textures, such as oyster meringue, one of elBulli's signature dishes.

**Sommelier**—A wine steward, responsible for helping diners select the appropriate wines and other beverages to complement their food. The sommelier has associated responsibilities as well, such as ordering and storing the wine.

**Spherification**—A cooking technique invented by Ferran Adrià of elBulli, in which a cooking liquid is reshaped into a sphere that resembles caviar.

**Terroir**—The total impact of a given location on a wine, including soil, elevation, and climate. Sometimes wine lovers translate this term as "somewhereness."

**SGN**—From the French *Sélection de Grains Nobles* (selection of noble berries), these are sweet wines, primarily from Alsace, made from grapes affected by noble rot (*botrytis cinerea*).

**Sous-voile**—A winemaking style in which the wine is intentionally exposed to oxygen. Casks are filled five-sixths full, leaving "the space of two fists" empty at the top, in order to allow a beneficial veil (or flor) of yeast to develop on top of the wine.

**VDN**—From the French *Vin Doux Naturel* (naturally sweet wine), these are fortified sweet wines typically made from Muscat or Grenache grapes. Fermentation is stopped by adding a grape spirit, producing a wine with an alcohol level between about 15 and 18 per cent.

# BIBLIOGRAPHY

## BOOKS

ADRIÀ, Ferran, Juli Soler, and Albert Adrià. *elBulli 1994-1997: A Period That Marked the Future of Our Cuisine*, HarperCollins Canada, 2006.

ADRIÀ, Ferran, Juli Soler, and Albert Adrià. *elBulli 2003-2004*, HarperCollins Canada, 2006.

BÉLIVEAU, Richard, Ph.D. and Denis Gingras. *Foods that Fight Cancer*, McClelland & Stewart, 2006.

BURDOCK, George A., Ph.D. *Fenaroli's Handbook of Flavor Ingredients, Fifth Edition*, CRC Press, 2005.

CHANG, Raymond. *General Chemistry*, McGraw-Hill, 1998.

CHARTIER, François. *À table avec François Chartier,* Les Éditions La Presse, 2005.

CLARKE, R.J. and J. Bakker. *Wine Flavour Chemistry*, Blackwell Publishing, 2004.

FLANZY, Claude. *Œnologie: fondements scientifiques et technologiques*, Lavoisier, 1998.

FRISQUE-HESBAIN, Anne-Marie. *Introduction à la chimie organique*, Hart/Conia, 2000.

FUNDACIÓ Alícia and elBullitaller. *Léxico científico gastronómico*, Planeta, 2006.

HILLS, Phillip. *Appreciating Wine*, Harper Collins, 2004.

McGEE, Harold. *On Food and Cooking: The Science and Lore of the Kitchen*, Scribner, 2004.

RICHARDS, J.H., D.J. Cram, G. R. Hammond, and P. L'Écuyer. *Elements of Organic Chemistry*, McGraw-Hill, 1968.

THIS, Hervé. *Molecular Gastronomy*, University of Columbia Press, 2005.

THIS, Hervé. *Traité élémentaire de cuisine*, Belin, 2002.

WATERHOUSE, Andrew L. and Susan E. Ebeler. *Chemistry of Wine Flavor*, ACS Symposium Series no 714, 1999.

## SCIENTIFIC PUBLICATIONS

CUTZACH-BILLARD. "Études sur l'arôme des vins doux naturels non muscatés au cours de leur élevage et de leur vieillissement. Son origine. Sa formation," Université de Bordeaux, 2000.

# INDEX